ELEMENTARY MEDICAL BIOPHYSICS

To

our

Students

ELEMENTARY MEDICAL BIOPHYSICS

by

G. G. JÁROS
M.Sc. (Eng.) (Wits.), M.(S.A.)I.E.E.
Lecturer in the Department of Physiology and Physiological Chemistry, University of Pretoria

and

B. J. MEYER
M.Sc. (Stell.), M.D., D.Sc., (Pret.)
Professor and Head of the Department of Physiology and Physiological Chemistry, University of Pretoria

DURBAN 1972

ISBN 0 409 08271 6

Reprinted 1980

THE BUTTERWORTH GROUP

South Africa
BUTTERWORTH & CO (SA) (PTY) LTD
152-154 Gale Street, Durban 4001

England
BUTTERWORTH & CO (PUBLISHERS) LTD
88 Kingsway, London, WC2B 6AB

Australia
BUTTERWORTHS PTY LTD
586 Pacific Highway, Chatswood, Sydney NSW 2067

Canada
BUTTERWORTH & CO (CANADA) LTD
2265 Midland Avenue, Scarborough, Ontario, M1P 4S1

New Zealand
BUTTERWORTHS OF NEW ZEALAND LTD
T & W Young Building, 77-85 Customhouse Quay, Wellington

USA
BUTTERWORTH PUBLISHERS INC
10 Tower Office Park, Woburn, MA 01801

Printed in South Africa by
INTERPRINT (PTY) LIMITED
DURBAN

Foreword

It is with pleasure that I comply with the request of the publishers to write a foreword to the textbook "Elementary Medical Biophysics".

The one author has had many years of experience in the planning and presentation of courses in physiology and biophysics for nurses and paramedical students, and the other has been lecturing in these subjects to this category of students for several years. The material presented is intended to cater for the particular needs of these students and has been thoroughly tested in teaching programmes.

As there is a great need for this type of basic textbook amongst tutors in nursing schools and in schools for paramedical staff in connection with the teaching of biophysics as an applied subject as against classical theoretical physics, the hope is expressed that this textbook will receive the general recognition it deserves.

The book is presented in the hope that it will supplement the training of nurses and paramedical staff in order that the life of the patient will at all times be in safe hands.

CHARLOTTE SEARLE
D.PHIL., R.G.N., R.M., DIP.N.E., D.H.A.
Professor and Head of the Department of Nursing
University of Pretoria

Pretoria
October 1972

Introduction

The spectacular development in the field of technology over the past decade or two has also had a marked influence on medical science.

This influence not only affects the medical researcher, medical student and general practitioner, but also the nurse and indeed all persons concerned with paramedical disciplines. To provide for the needs of our nursing students, an elementary course in medical biophysics was introduced. Generally speaking, the course met with so much approval that it was decided to make it available in book form. As the title suggests, it is no more than an elementary description of the most important biophysical concepts. We realise only too well that attempting to keep the course elementary may lead to scientific misconceptions. However, we trust that the positive value of our efforts outweighs the shortcomings.

We express our sincere thanks to Professors J. G. Koen (Department of Physics), L. van Biljon (Department of Electrotechnical Engineering) and J. M. M. Brown (Department of Physiology and Physiological Chemistry) all of the University of Pretoria, and Dr. M. Milner (Department of Bio-Engineering and Medical Physics) of the University of Cape Town, for their meaningful advice.
A special word of thanks to Mrs. K. Járos who was responsible for the design and execution of the illustrations, to Mrs. J. Roché for reading and to Mrs. H. Botha for typing the manuscript.

G. G. Járos

B. J. Meyer

Pretoria

October, 1972

Contents

SECTION II — ELECTRICITY

Chapter 4 — The Origin of Electricity and the Flow of Electricity in Conductors

Chapter 5 — Production of Electricity in Electrical and Living Cells

Chapter 6 — Variation and Measurement of Electrical Currents

Chapter 7 — Bio-Electronic Equipment

SECTION III — LIQUIDS, GASES AND HEAT

Chapter 8 — Common Properties of Liquids, Gases and Heat

Chapter 9 — Gases and Respiration

Chapter 10 — Liquids and the Circulatory System

Chapter 11 — Fluid Mixtures

Chapter 12 — Heat and the Regulation of Body Temperature

SECTION IV — RAYS AND WAVES

Chapter 13 — Electromagnetic Radiation

Chapter 14 — Light and Sight

Chapter 15 — Radioactivity

Chapter 16 — Sound and Hearing

SECTION V — MECHANICS

Chapter 17 — Machines

Appendix A: The Elements and their Symbols

Appendix B: Examples of Examination Questions

SECTION I

BASIC PRINCIPLES

CHAPTER 1

Graphical and mathematical representation of scientific problems

Introduction

Physical phenomena are expressed in terms of *natural laws*. Most of these laws can be presented graphically. The practical value of this is that relationships between physical quantities can be represented in a clear and concise manner. Some of these laws can also be expressed in mathematical terms. Mathematical equations can be manipulated to obtain useful information about the physical phenomena in question.

The procedure followed when formulating a physical law involves the following steps:

1. general observation of the phenomenon;
2. experimental measurements;
3. tabulation of experimental results;
4. construction of a graph;
5. formulation of a mathematical equation, if possible;
6. formulation of the law.

To demonstrate this procedure, the relationship between heat energy and temperature can be used as an example. We all know that in most instances, when heat is conducted to an object, its temperature will rise. As it stands, this observation is qualitative. Physics, however, is a quantitative science, and in order to formulate a law on the basis of the above-mentioned observation it must be shown that the temperature for a fixed amount of heat energy will rise by a fixed amount. The particulars can be obtained experimentally as follows: a beaker containing water is placed on an electrical stove, measurements are made to establish by how much the temperature of the water rises for different quantities of heat energy absorbed. The results can be tabulated as shown in TABLE 1.1 and can be used further to obtain a graph (FIG. 1.1).

Heat energy (energy units)	*Temperature (temperature units, e.g. °C)*
0	0
4	1
8	2
12	3
24	6
36	9

TABLE 1.1 Results of an experiment in which the relationship between temperature and heat energy was studied.

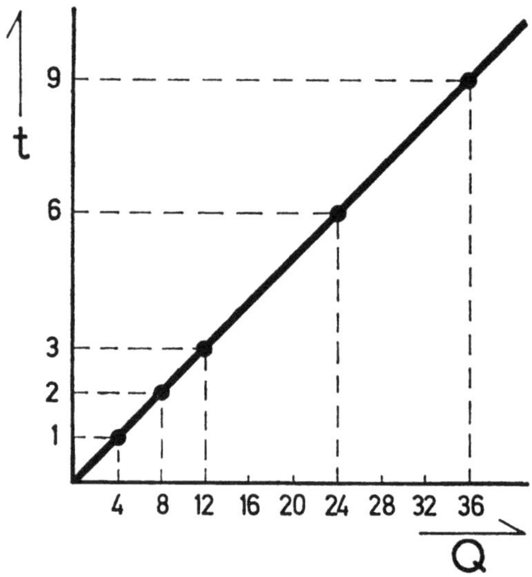

FIG. 1.1 Graphical representation of the straight line relationship between heat energy and temperature. t = temperature in degrees Celsius (°C); Q = heat energy in energy units.

The independent variable (the cause of the phenomenon) is plotted on the horizontal axis of the graph and on the vertical axis, the dependent variable (the result of the phenomenon). In our example, the temperature (t) is the dependent variable and the heat-energy (Q) the independent variable. The independent variable is responsible for the phenomenon, i.e. the temperature is dependent on the heat energy supplied to the object and not *vice versa*.

The form of the graph in this case is a straight line, but it could have had any other form. For example, a graph that illustrates the variations in a person's blood glucose concentration with time after consuming 50 g sugar (Glucose Tolerance Curve), can take different forms (Fig. 1.2).

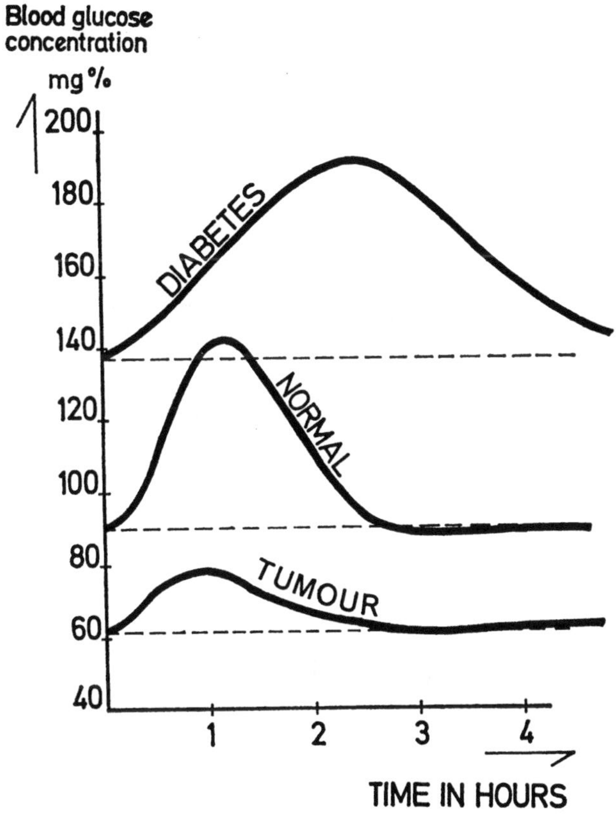

Fig. 1.2 Glucose Tolerance Curve. A graphical representation of how the glucose concentration in blood changes with time after 50 g glucose is ingested. Blood samples are taken just before intake (t = 0) and hourly thereafter. The three graphs represent the changes in glucose concentration of a normal person, of a diabetic person and a person with a tumour in the island tissue of Langerhans. After A. C. Guyton, Textbook of Medical Physiology, 1966.

The above-mentioned figure includes three graphs, viz, one representing results obtained from a healthy person and the other two representing the results from persons with pancreatic malfunction. The form of the curve for the normal person differs from that obtained from persons suffering from diabetes mellitus and from those persons with a tumour in the islands of Lagerhans in the pancreas. The value of a graph is evident from the Glucose Tolerance Curve. In a normal fasting person, the blood glucose concentration is about 90 mg/100 ml blood. When 50 g of glucose is administered, the concentration of glucose in the blood increases to about 140 mg/100 ml blood within the first hour, but returns to the normal value after about 2½ hours. In a patient with diabetes mellitus, the resting value for the glucose concentration is higher (140 mg/100 ml). When the 50 g glucose is ingested the glucose concentration in the blood rises slowly over a period of 3 hours to 200 mg/100 ml or higher and then returns gradually to normal. A patient with a tumour in the islands of Langerhans in the pancreas has a fasting blood glucose concentration of 60 mg/100 ml. The administration of glucose in this case only increases the concentration to 80 mg/100 ml. Notice how many words were used to describe graphs, and thus how much time is saved by using a graph. Mathematical analysis of the Glucose Tolerance Curve is complicated and for our present purpose unnecessary.

In Physics, most of the simple laws can be explained by means of graphs and mathematical analyses. In this chapter the following types of graphical representation will be discussed briefly: the straight line, the hyperbola, the parabola and the inverse square equation.

Straight line

When the relationship between two variables can be represented by a straight line, we talk about proportionality between the two variables. *Proportionality* means that the percentage change in one of the variables corresponds to the percentage change in the other variable. For example, from FIG. 1.1 it can be seen that 4 units of heat energy increase the temperature by 1°C, 8 units by 2°C (both have doubled) and 12 units by 3°C (both have trebled). It is evident from the graph that for every unit change in heat energy the temperature changes by a constant amount (i.e. 1/4°C). This means that at every point on the graph

$$t = 1/4\,Q$$

The constant value (1/4) thus represents the change in the dependent variable for each unit change in the independent variable. The above equation applies to the straight line obtained in a specific experiment. The general form used to represent a straight line is

$$t = kQ$$

where k is a constant that varies from experiment to experiment and is equal to the slope of the straight line, i.e. the constant equals the height increment for each unit increase in the horizontal direction. With the aid of the above equation, more particulars can be obtained concerning the physical phenomenon without any additional measurements. For example, the amount of heat energy necessary to raise the temperature of water from 0°C to 100°C in the above-mentioned experiment, can be determined by rewriting the equation thus:

$$Q = \frac{t}{k}$$

$$= \frac{100}{0\ 25} = 400 \text{ units}$$

After repeating the experiment with various substances and finding that they react in the same manner, the final formulation of the law may read as follows: the temperature of an object increases proportionally with the amount of heat energy supplied, or simply

$$t = kQ$$

where t = temperature and Q = heat energy

Limitations of physical laws

It is important to remember that physical laws have certain limitations, i.e., they do not in all circumstances provide the right answers. The above-mentioned law does not for example, apply at the melting point and boiling point of the material (FIG. 1.3), as at

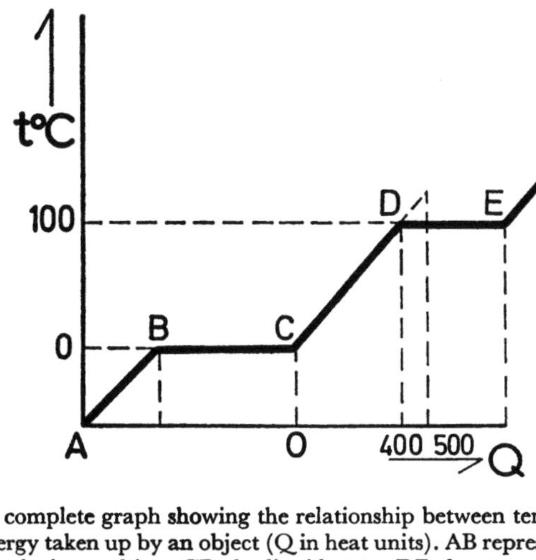

FIG. 1.3 A complete graph showing the relationship between temperature (t in °C) and heat energy taken up by an object (Q in heat units). AB represents the solid state, BC the state during melting, CD the liquid state, DE the state during evaporation and EF the gaseous state.

these points heating does not increase the temperature. The heat energy supplied at the melting point to solids is used to change their structure in such a way that they become liquids, and the heat energy supplied to a liquid at the boiling point changes the liquid into a gas. These phenomena will be discussed further in Chapters 8 and 12. When a physical law is formulated, its limitations must always be taken into account. The above-mentioned law in connection with heat and temperature should therefore be stated as follows: *the temperature of a material* changes proportionally with the amount of heat energy supplied to it, except at the melting and boiling points.* Physical laws are often used beyond the confines of their limitations giving rise to incorrect results. This can be illustrated with the above-mentioned physical law where $t = 1/4\ Q$. Let us suppose that we want to determine the temperature of an object, initially at 0°C, after 500 units of heat energy have been supplied to it. If we use the equation $t = kQ$, the result will be $t = 1/4 \times 500 = 125$°C. From FIG. 1.3, it is obvious that the temperature is still at 100°C because the water is in the process of evaporating.

The limitations of a physical law are as important to remember as the law itself, since if one is unable to remember the law itself the answer to a problem cannot be determined, but if the law is remembered and incorrectly applied the conclusions may be erroneous. Another example illustrating this point, is Starling's law of the heart. According to this law, the heart "under normal physiological conditions pumps all the blood that it receives". The limits in this case are set by the speed at which blood reaches the heart under normal circumstances. If excessive amounts of blood reach the heart it will not be able to pump all this blood and the end result will be heart failure.

Hyperbola

A phenomenon that can be represented graphically by a hyperbola (FIG. 1.4), is the example which can be described as follows: "*when*

V	*P*	*P* × *V*
1	5	5
2	2,5	5
5	1	5
10	0,5	5

TABLE 1.2 Results of an experiment in which the relationship between the pressure (P) and the volume (V) of a gas was studied. The third column shows that the product of P and V remains constant throughout experiment.

* There are certain materials that are exceptions to this rule.

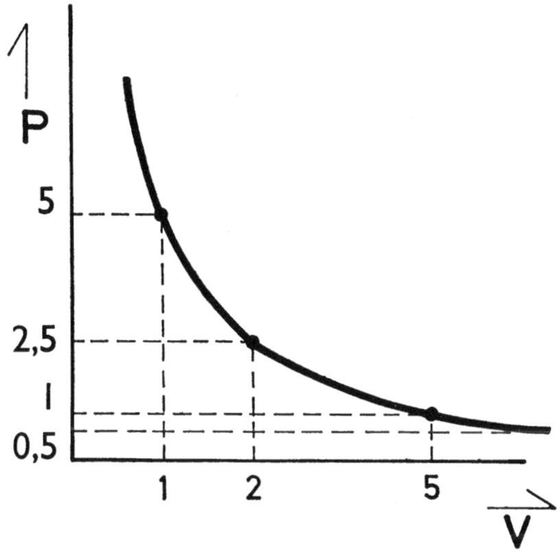

FIG. 1.4 Graphical representation of the hyperbolic relationship between the pressure (P) and volume (V) of a gas.

a gas in a closed container is compressed, its pressure increases". Experiments can be performed by enclosing a constant amount of gas in containers having different volumes (V), and at the same time measuring the pressures (P) in the containers (TABLE 1.2).
It is evident from the third column of TABLE 1.2 that the product of the two variables is constant. Mathematically the relationship can therefore be written as $P \times V = k$. In the above experiment $k = 5$. Rewriting the equation as, $P = \frac{k}{V}$, one can see that as V increases P decreases and *vice versa*, or more precisely when V is doubled P becomes halved. The equation represents inverse proportionality. The above-mentioned law in connection with the volume and pressure of gases will be discussed again in Chapter 9.

Parabola

It is well known that the surface area of a circle does not increase proportionally to the radius but increases to a *much greater extent* as the circle is expanded. To show that this really is the case we can measure the surface areas of circles of various radii and tabulate the results (TABLE 1.3). It is clear that the percentage increase in the surface area exceeds the percentage increase in the radius. The measured values can also be plotted out in the form of a graph (FIG. 1.5). The graph obtained is called a parabola. The mathematical expression for a parabola is, $A = kr^2$.

r (m)	*A* (m^2)
1	3,14
2	12,48
3	28,26
4	50,24

TABLE 1.3 Results of measurements performed in order to establish the relationship between radius (r) and the area (A) of a circle.

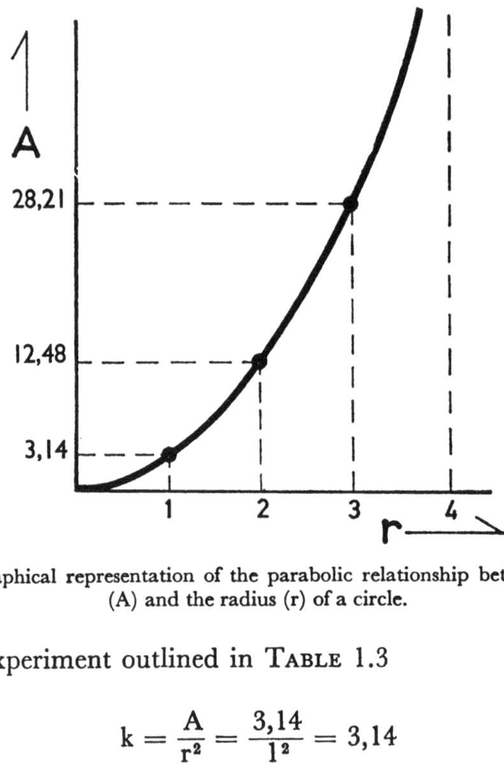

FIG. 1.5 Graphical representation of the parabolic relationship between the area (A) and the radius (r) of a circle.

From the experiment outlined in TABLE 1.3

$$k = \frac{A}{r^2} = \frac{3,14}{1^2} = 3,14$$

The Greek letter π is used to symbolise the constant 3,14. The surface area of a circle can thus be expressed as

$$A = \pi r^2$$

This parabolic equation means that A is dependent on changes involving the square of the value of r and not the value of r itself. For example if r is doubled (2 × r), A increases fourfold ($2^2 = 4$), and if r increases 5 times (5 × r), A increases 25 times ($5^2 = 25$).

Inverse square equation

It has already been shown that two quantities can be inversely proportional to each other. In the same way an inverse square relationship is possible (TABLE 1.4 and FIG. 1.6). The graph obtained is similar to a hyperbola, except that the dependent variable (Y) decreases to a greater extent with an increase of the independent variable (Z) than in the case of the hyperbola. Mathematically the relationship can be written as follows:

$$Y = \frac{k}{Z^2}$$

and with k = 4 in the above example

$$Y = \frac{4}{Z^2}$$

Z	Y
1	4
2	1
4	$\frac{1}{4}$

TABLE 1.4 Results of an experiment in which the relationship between two quantities, Z and Y, is studied.

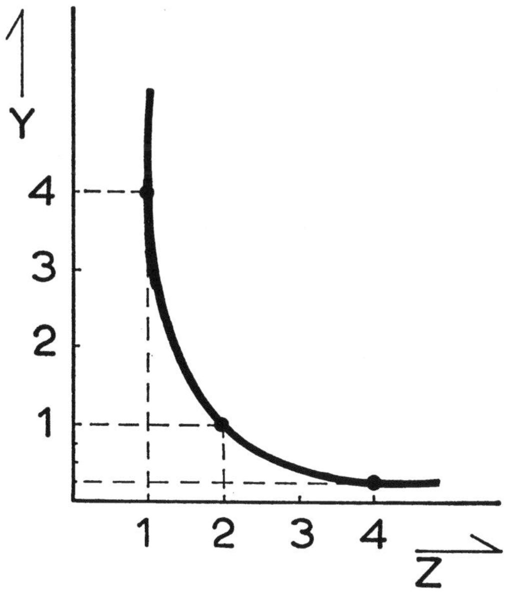

FIG. 1.6 Graphical representation of the inverse square relationship between two quantities, Y and Z.

Equations with more than two variables

It has been mentioned that the relationship between the pressure (P) and volume (V) of a gas can be expressed by a hyperbola, i.e. $P = k_1/V$, where k_1 is a constant for the specific experiment. It has also been explained that each physical law has limitations and that it is necessary to keep these limitations in mind. The above-mentioned law for instance, only applies if the number of molecules of the gas and the temperature of the gas are kept constant. If, however, the temperature (T) and the volume (V) are kept constant but the number of molecules (n) changes, the following expression $P = k_2\,n$ applies to the relationship between the pressure (P) and the number of molecules (n). If the volume (V) and the number of molecules (n) are kept constant, the relationship between pressure (P) and the temperature (T) becomes $P = k_3\,T$.*

The relationship between the pressure, the volume, the temperature and the number of molecules can thus be expressed by three equations. These three equations can be combined to form a single equation as follows:

$$P = \frac{KnT}{V}$$

In this case all the above-mentioned limitations are included in the equation. When n and T are constant, the equation becomes

$$P = \frac{(K \times n \times T)}{V} = \frac{k_1}{V}$$

$$\text{where } k_1 = K \times n \times T$$

When T and V are constant the equation becomes

$$P = \frac{K \times T}{V} \times n = k_2 n$$

$$\text{where } k_2 = \frac{K \times T}{V}$$

When n and V are constant the equation becomes

$$P = \frac{K \times n}{V} \times T = k_3 T$$

$$\text{where } k = \frac{K \times n}{V}$$

To interpret equations with more than two variables it is therefore necessary that the variables be taken in pairs, while the other variables are kept constant.

* T is measured in absolute units, i.e. in kelvin (K).

CHAPTER 2

Units

Introduction

In Physics, quantities, such as distance, time and energy are measured when laws are to be formulated, or when particulars of known laws are investigated. To express quantities, units are necessary. For example, it is meaningless if we say that the distance between Johannesburg and Cape Town is 1 600 without adding the units in which the distance is expressed. There are naturally exceptions to the rule. It can be said that the red blood cell count is 5 million or that the haematocrit is 35 without mentioning the units. In these cases the units are omitted because the people concerned should know that the red blood cell count is always expressed as cells per cubic millimetre (microlitre) and the haematocrit as volume percent.

Quantity	*Unit*	*Symbol*
Length	metre	m
Mass	kilogram	kg
Time	second	s
Electrical current	ampere	A
Thermodynamic temperature	kelvin	K
Light intensity	candela	cd

TABLE 2.1 The six fundamental units of the SI.

The SI units

Measurement means expressing dimensions in terms of known units. If we say that a ruler is 300 mm long we mean that its length is 300 times that of the known unit, in this case the millimetre. The emphasis is on the word *known*. If we talk about the millimetre people all over the world know what we mean. The standard unit of length is the metre. This standard was established in 1781 and is kept in the form of a rod made of 90% platinum and 10% iridium at the Bureau of Standards in Paris. The metre was defined then as the distance between two marks made on this rod at a temperature of 0°C. The modern definition is considerably more complex and need not concern us at the moment. Accurate copies of this particular metre rod are to be found in most capital cities in the world. Sets of standards are also available in most large centres and these are periodically checked against national and international standards.

At the Conférence Générale des Poids et Mesures, that took place in Paris in 1960, standards were adopted for mass, time, electrical current, thermodynamic temperature and light intensity (TABLE 2.1). Along with length, these quantities are called the fundamental quantities. This system of fundamental units and its derivatives (TABLES 2.2 and 2.3) is called the "Systéme International" (SI), (International System).

Derived units of the SI

Derived units can be divided into two groups, viz, those with special names and those with complex names. Units which are in general use are summarized in TABLES 2.2 and 2.3.

Quantity	*SI unit*	*Symbol*	*Derivation*
Frequency	hertz	Hz	t^{-1}
Force	newton	N	kgm/s^2
Work, energy, quantity of heat	joule	J	Nm
Power	watt	W	J/s
Electrical Charge	coulomb	C	As
Electrical Potential	volt	V	W/A
Electrical Capacitance	farad	F	C/V
Electrical Resistance	ohm	Ω	V/A
Pressure	pascal	Pa	N/m^2

TABLE 2.2 Derived units of the SI with special names.

Quantity	*SI unit*	*Symbol*
Surface Area	square metre	m^2
Volume*	cubic metre	m^3
Density	kilogram/cubic metre	kg/m^3
Velocity (speed)	metre/second	m/s
Acceleration	metre/second/second	m/s^2

TABLE 2.3 Derived units of the SI with complex names.

* The litre is used for measurement of fluids: 1 litre = 0,001 m^3

Multiples and submultiples of the SI units

It is not always practical to use the above-mentioned units. For example, if it is the thickness of paper that is to be measured, the answer could be, say, 0,000 1 metre. On the other hand the distance between Johannesburg and Cape Town is 1 600 000 metre. To solve the problem of inconveniently long numbers, multiples and submultiples of units have been introduced. The multiples and submultiples differ from one another in powers of ten. For this reason the SI is sometimes called the decimal system (deci = tenth). To express multiples and submultiples prefixes with Greek and Latin names are used (TABLE 2.4).

Prefix	*Symbol*	*Factor with which the unit must be multiplied*
Tera	T	10^{12}
Giga	G	10^{9}
Mega	M	10^{6}
kilo	k	10^{3}
milli	m	10^{-3}
mikro	μ	10^{-6}
nano	n	10^{-9}
piko	p	10^{-12}

TABLE 2.4 Prefixes to express multiples and submultiples.

N.B.: (1) $10^{-3} = \frac{1}{10^3}$ en $10^{-12} = \frac{1}{10^{12}}$

(2) In the SI the prefixes centi (10^{-2}), deci (10^{-1}) and hecto (10^{2}) are only used in a few special instances.

Conversion of subunits

To convert from one subunit to another the factors listed in TABLE 2.4 are used.

Examples

1. Calculate how many millimetres there are in 0,035 km

$$\begin{aligned} 0{,}035\ (\text{k})\text{m} &= 0{,}035 \times (10^3)\ \text{m} \\ \text{and } 10^3\ \text{mm} &= 1\ \text{m} \\ \therefore\quad 0{,}035 \times 10^3\ (\text{m}) &= 0{,}035 \times 10^3 \times (10^3\ \text{mm}) \\ &= 0{,}035 \times 10^6\ \text{mm} \\ &= 35 \times 10^3\ \text{mm} \end{aligned}$$

2. Calculate how many metres there are in 1 km^2

$$1\ \text{km}^2 = 1 \times (10^3\ \text{m})^2 = 10^6\ \text{m}^2$$

3. Calculate how many pm^3 there are in 0,5 Gm^3 (Gm = gigametre)

$$\begin{aligned} 1\ \text{Gm} &= 10^9\ \text{m} \\ 1\ \text{m} &= 10^{12}\ \text{pm} \\ \therefore\quad 1\ \text{Gm} &= 10^9 \times 10^{12}\ \text{pm} = 10^{21}\ \text{pm} \\ \therefore\quad 0{,}5\ \text{Gm}^3 &= 0{,}5\ (10^{21}\ \text{pm})^3 \\ &= 0{,}5 \times 10^{63}\ \text{pm}^3 \end{aligned}$$

Rules for the usage and writing of the SI

1. A comma must separate the integers from the decimal fractors of numbers.

   e.g. 1,2 or 15,35

2. Numbers consisting of many figures must be separated in groups of three counting left and right from the decimal comma. The groups must be divided by a space, and never by comma or point:

   e.g., 2 655 425,725 424

3. In numbers smaller than one the decimal comma must be preceded by a zero, e.g., 0,239 and not ,239.

4. Symbols for units must only begin with capital letters if the symbol is derived from the name of a person. If, however, the same unit is written out in full it must commence with a small letter,

   e.g. newton, N; hertz, Hz.

5. The rules for units and symbols must be observed at all times,

   e.g. gram — g and not gr or gm
   kilogram — kg and not Kg or Kgm

6. Prefixes and units are written as one word.

   e.g. millimetre, microgram

7. Complex prefixes such as millimicro are to be avoided.

8. When using symbols, no addition such as "s" should be made to indicate plurality, 2 kg and not 2 kgs.

9. Numerical values lying between 0,1 and 1 000 are very convenient in denoting dimensions. The appropriate prefixes must be selected.

   e.g. 17 000 m = 17 km
   0,002 56 kg = 2,56 g

Calibration

Calibration means comparing the readings of an unknown instrument with the readings of a standard instrument. If the two agree completely the unknown instrument is a good instrument. If the readings on the unknown instrument do not agree with those of the standard completely, the new instrument can still be used provided that the necessary corrections are made to its readings. To do this a graph or table of corrections (calibration curve or table) must be constructed. In the case of an uncalibrated ruler, it may be calibrated by holding it against a calibrated ruler and reading off the corresponding values. These values are then tabulated (TABLE 2.5). The table can be used to construct a correction curve (FIG. 2.1).

Unknown	*Standard*	*Correction factor*
0	0	0
10	11	+1
20	22	+2
30	32	+2
40	43	+3
50	53	+3
60	65	+5
70	72	+2
80	82	+2
90	93	+3

TABLE 2.5 Correction Table of Ruler

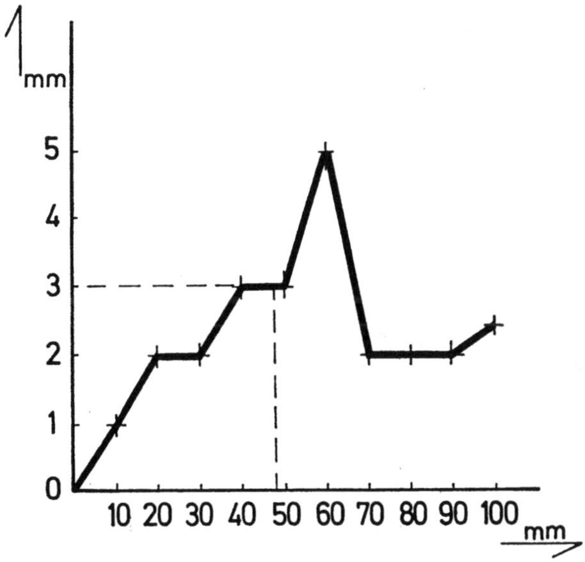

FIG. 2.1 Correction curve for the ruler from Table 2.5. The figures on the horizontal axis represent the reading on the new ruler and the figures on the vertical axis the correction factors.

The following example illustrates the use of the correction curve. Assume that the length read from the new ruler in the above example is 48 mm, find its correct length.

From the table or graph the correction factor = 3 mm
∴ Correct length = 48 + 3 = 51 mm

Errors in measurement

When we measure something the result is not always completely accurate since errors may occur. Two kinds of errors can be made, viz.,

(i) *constant errors.* This type of error has already been demonstrated in the case of the ruler. If an instrument is calibrated wrongly, constant errors can occur. These errors can be eliminated by recalibrating the instrument.

(ii) *Incidental errors.* These are errors arising in the manner of reading, for instance when a number of readings are made by the same person or by a different person each time. They relate to the human factor rather than to the instrument itself.

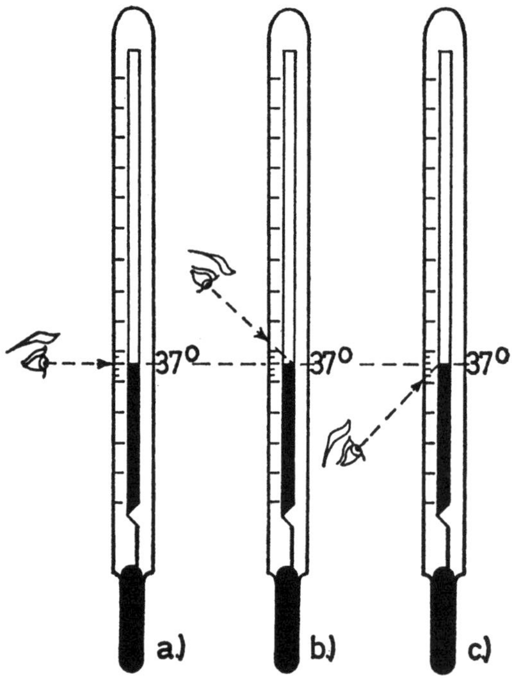

FIG. 2.2 Different methods of reading a thermometer. Method (a) is correct; (b) and (c) are incorrect.

If a thermometer for example is read with the eye at a level above the mercury meniscus in the tube, a higher value will be obtained than when the eye-level is below the meniscus (FIG. 2.2). This error can be eliminated by repeating the measurements and taking the average, or by simply keeping the levels of the eye and mercury at the same height as is usually done.

Measurement of temperature

Temperature is measured by means of a thermometer. The principles of measurement are based on the fact that materials* expand when they are heated and contract when cooled. Mercury or alcohol is generally used in thermometers. The mercury or other liquid in the tube of the thermometer occupies a certain length of the tube at each temperature. This makes it possible to mark off standard points in the length of the tube which coincide with the length of the mercury column at certain temperatures. The mercury column will be at a certain level at the freezing point of water and at another level at the boiling point of water. Celsius called the former 0° and the latter 100°, and divided the distance between the two into 100 equal parts, each of which represents 1° Celsius (1°C).

The absolute temperature scale is also used in science. The absolute unit of temperature is the K (kelvin). The unit has the same magnitude as the °C. The difference between the two systems is that the zero point of the absolute system coincides with —273°C. In other words, the freezing point of water on the absolute scale is 273 K, the boiling point 373 K and the normal temperature of the human body is 273 + 37 = 310K.

*There are certain materials that are exceptions to this rule, e.g. water between 0° and +4°C.

CHAPTER 3

Forces

Introduction

Physics is the science in which the influence of forces on materials is investigated. According to this definition, forces play an active role and materials a passive role. Force can be defined as the influence which changes or attempts to change the equilibrium states of materials. Resting states and motion at constant speed in a straight line are considered to be equilibrium states. Movement of parts of materials with respect to one another is known as deformation or change of form. The above-mentioned definition of force is actually another way of expressing the first law (N_1)* of Newton which he formulated three centuries ago. According to N_1 a body will remain at rest or in motion in a straight line at a constant speed until external forces act upon it.

Natural processes depend on physical or chemical laws involving some type of force. It is often diffcult to define the force or state how it arises, yet the effects of the force, for instance in the change of equilibrium, or efforts to change this, are clearly visible.

Generally, forces work in pairs, or as it is stated in N_3, for every action there is a reaction or for every force an opposing force. This means that there is generally equilibrium between two forces acting on an object. Where one of the forces exceeds the other, the resting state or state of motion changes until a new equilibrium is attained. In nature there is a strong tendency towards equilibrium, which means stability, since change of state means instability. For our purposes it is sufficient to accept the fact that forces exist, without going into further details about the reasons for their existence or their origins.

Gravitational force

Gravitational force is the force of attraction that exists between two objects. There is a force of attraction between the sun and its planets, the earth and the moon and the earth and objects on the surface of the earth. The common name for this force which is exerted by the earth on an object is the *weight* of the object.

* N_1, N_2 and N_3 are abbreviations for the three laws of Newton.

The size of the gravitational force is given by the following formula:

$$F_G = K_G \frac{M_1 M_2}{d^2}$$

F_G = the gravitational force in newton
M_1 and M_2 = the masses of the objects in kg
d = the distance between the objects in m
K_G = the gravitational constant

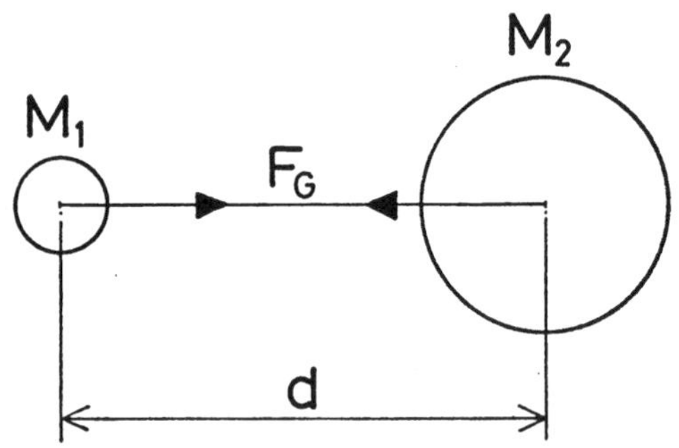

FIG. 3.1 Schematic representation of the gravitational force (F_G) that exists between two objects. M_1 and M_2 = masses of the two objects and d = the distance between the two objects.

Mass (M) is the quantity of material that is present in an object and is dependent upon the structure of the material of the object. The structure is determined by the size and concentration of the atoms making up the material. Every material has its own density which indicates the mass of the material per unit volume. This means that the mass (M) of an object can be calculated by multiplying the volume (V) with the density (ρ) i.e. $M = \rho V$.
According to the above explanation the *mass* is the quantity of material in the object and it is the same everywhere in the universe. Weight, however, is a completely different concept that can be expressed by the following equation:

$$W = F_G = K_G \frac{M_E M_O}{d^2}$$

F_G = the force of attraction exerted on the object on the surface of the earth by the earth in newton
M_E = mass of the earth in kg = constant

M_O = mass of the object in kg
K_G = constant
d = radius of the earth in m. The radius of the earth changes from place to place as the height above sea level changes, but it is constant at any given place on earth

$\therefore W = g\,M_O$ where $g = \frac{K_G M_E}{d^2}$ is constant for a certain place on the surface of the earth.

According to this equation the weight of an object changes from place to place on earth, yet remains the same at any given place on earth. When a person picks up an object from the ground he exerts a force which is greater than the gravitational force exerted upon the object by the earth. The life of man on earth is to a large extent determined by the gravitational force. His muscles exercise a certain amount of force which is required to hold him upright despite the effect of his weight. If the muscular effort is not exerted, as happens when a person loses consciousness, he simply falls down. When a person moves an arm or a leg this is also done against the gravitational force of the earth.

Centrifugal force

Centrifugal force exists whenever an object moves in a circle (FIG. 3.2). This force tends to compel the object away from the centre of movement. If an object is fixed to the end of a string and rotated around a central point it will shoot out at a tangent to the circle if the string is suddenly cut or if it breaks.
The magnitude of the centrifugal force (F_G) is given by the following equation

$$F_S = K_S \frac{Mv^2}{d}$$

F_S = centrifugal force in newton
M = the mass of the moving object in kg
v = the speed with which the object moves in m/s
d = the distance of the object from the centre of movement in m
K_S = a constant.

An example of the application of the principle of the centrifugal force in clinical medicine is the centrifuge. This apparatus is used for instance to separate blood cells from the liquid portion (plasma) in blood.

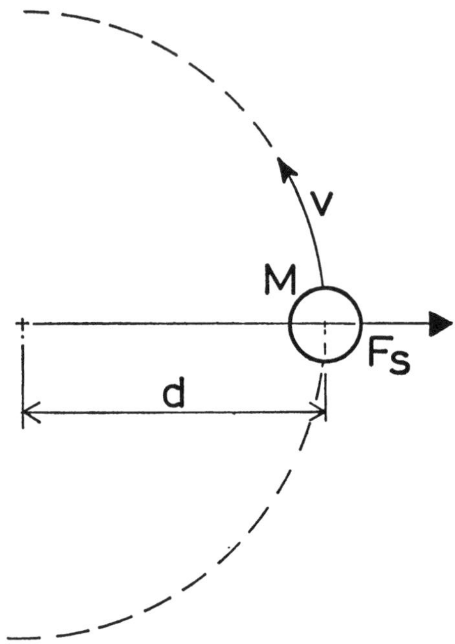

FIG. 3.2 Diagrammatic representation of the centrifugal force (Fs) acting on an object that moves in a circle. M = mass of the object, v = speed with which the object is travelling, d = radius of the circle.

The procedure is as follows. A few ml blood are placed in a calibrated test tube. It is placed in the centrifuge, which simply rotates the tube around a central point at great speed, in such a way that the bottom of the tube adopts a position furthest from the centre of rotation. During centrifugation the particles in the blood are acted upon by the centrifugal force. The force is proportional to the mass and therefore the cells which have the greatest mass will move furthest from the centre, namely to the bottom of the tube. In this way the cells and the liquid part of blood become separated. The volume occupied by the cells, expressed as percentage of the total volume of blood in the tube is called the haematocrit. Normally this value is in the vicinity of 45%, but varies in certain disease states. The determination of the haematocrit is therefore of great clinical value.

Equilibrium between the gravitational force and the centrifugal force

According to N_3, forces and opposing forces balance each other during a state of equilibrium. The moon moving around the earth is in equilibrium: it moves with a constant speed at a constant distance from the earth.

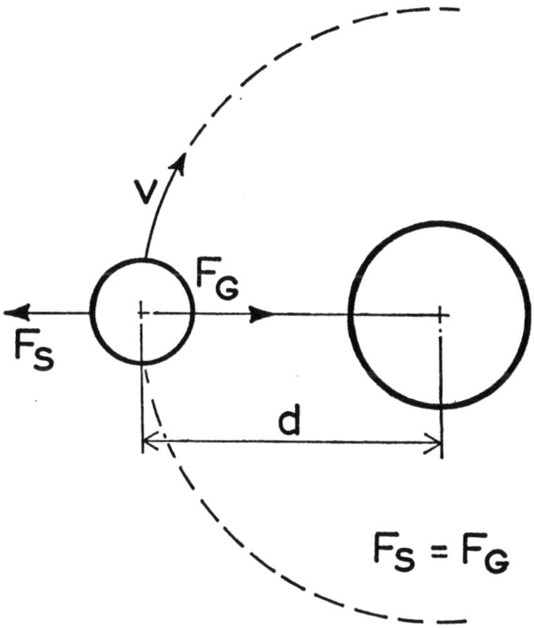

FIG. 3.3 Diagrammatic representation of the equilibrium between the gravitational force (F_G) and the centrifugal force (F_S) which act on an object rotating around another object. d = distance between the two objects, v = speed of motion.

Two equal forces are acting on the moon, viz, the gravitational force that attracts the moon to the earth and the centrifugal force which repels the moon from the earth. As a result of these two forces the moon does not move towards or away from the earth. In other words, there is equilibrium between the two forces, because they have equal magnitudes but work in exactly the opposite directions. This relationship can be written shortly as follows:

$$F_G = F_S$$

Electrical forces

Atoms of all materials contain positive protons and negative electrons. When two protons or two electrons are placed close to each other they repel one another. When a proton and an electron are placed close to each other they attract one another. The magnitude of the forces of attraction or repulsion is given by the formula

$$F_E = K_E \frac{Q_1 \times Q_2}{d^2}$$

F = the magnitude of the force in newton

Q_1 and Q_2 = the charge on the protons and electrons in coloumb

d = distance between proton and electron in m

K_E = a constant

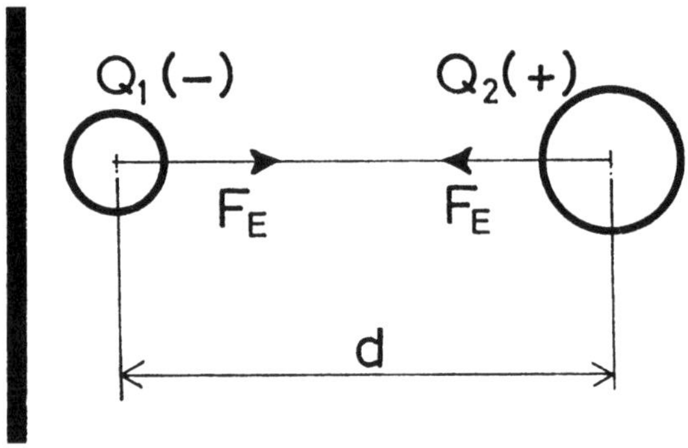

FIG. 3.4 Diagrammatic representation of the electrical force (F_E) between two charges. Q_1 and Q_2 = charges, d = distance between the charges.

The structure of atoms is very much like a miniature solar system, i.e. the electrons move in circular paths around the nucleus, just as the planets move around the sun. As is the case with the planets in the solar system, two forces act on the electrons.
These are best understood if we consider a simple atom, the hydrogen atom (FIG. 3.5). The hydrogen atom consists of one proton in the nucleus and only one electron which moves in a circular path around the nucleus. The force of attraction, the electrical force exerted by the two charges, is denoted by F_E, and the force of repulsion, the centrifugal force, by F_S. When $F_E = F_S$ the electron moves at a constant speed at a constant distance from the nucleus. Because the mass of the particles is very small, the gravitational force between them is negligible.

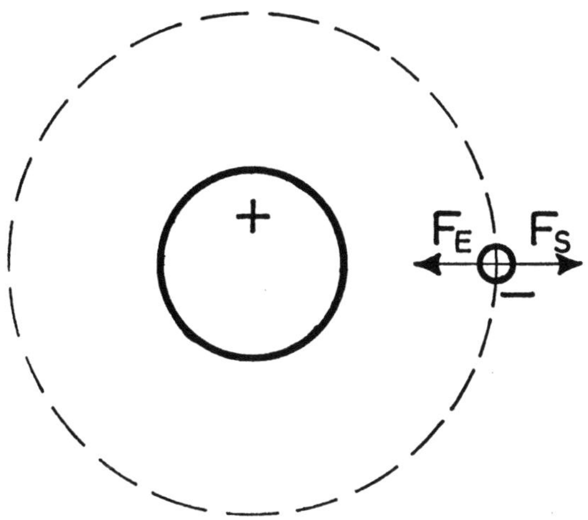

FIG. 3.5 Diagrammatic representation of the equilibrium between an electrical force (F_E) and a centrifugal force (F_S) when one charge moves around another.

Electrical forces play an important role in the human body. One example is afforded by muscle contraction. Every muscle consists of parallel muscle fibres which are built up of parallel muscle fibrils. The fibrils in turn consist of smaller units or segments (sarcomeres) which are placed in series along the length of the fibril. Each sarcomere consists of two types of filament-shaped protein molecules, i.e. actin and myosin, arranged in such a manner that their ends overlap (FIG. 3.6).

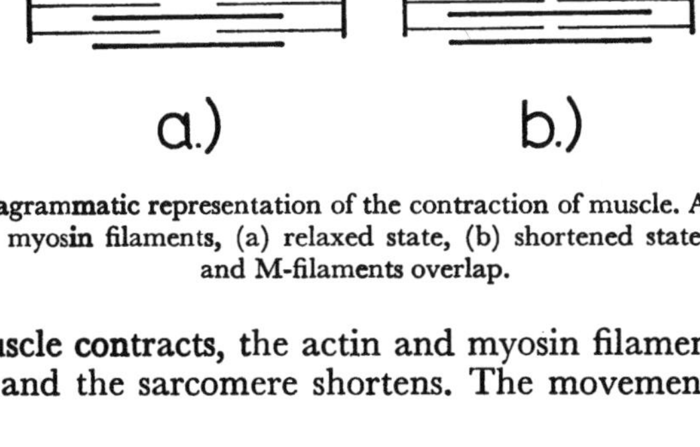

FIG. 3.6 Diagrammatic representation of the contraction of muscle. A = actin filaments, M = myosin filaments, (a) relaxed state, (b) shortened state. N.B. the A- and M-filaments overlap.

When a muscle contracts, the actin and myosin filaments slide over each other and the sarcomere shortens. The movement is initiated by electrical forces between the filaments which are produced before contraction. The myosin filaments become positively charged and the actin filaments negatively charged, and they thus attract one another. Although the shortening of each sarcomere is small, the total shortening of the muscle fibril is appreciable, since it is built up from a great number of sarcomeres.

Magnetic forces

Magnetic forces have been known for a very long time. Travellers have used magnets in the form of compasses since the early days of history to determine their direction and location. The ability to become magnetic, or to be attracted by magnets is an inherent property of certain metals.

The ends of two metal bars that are magnetic attract or repel each other, just as in the case of two electrical charges. If the ends or poles have the same kind of magnetism, they repel one another and if their magnetism differs, they attract each other. The magnitude of the magnetic force can be determined with the aid of the following formula:

$$F_M = K_M \frac{M_1 M_2}{d^2}$$

F_M = the magnetic force in newton

M_1 and M_2 = the strength of the poles in SI units

d= distance between the poles in m

K_M = a constant

A compass is a light weight magnet that is capable of turning around an axis. It can be used to determine direction since the earth itself is a great magnet, the South and North poles being the two magnetic poles. The compass magnet aligns itself in such a way that the pole which has magnetism opposite to that of the North pole will be nearest to the North pole, and that which has magnetism opposite to that of the South pole will be nearest to the South pole. The compass thus shows the N-S direction from which all other directions can be calculated.

Magnetism can be temporarily induced in metals by an electrical current flowing in a coil wound around the metal. It can even be induced in air or in a vacuum by such a coil. In the case of electrically induced magnetism, one speaks of electromagnetism, the principles of which are important in connection with measuring instruments, many of which are used in medicine. Electromagnetism is discussed again in Chapter 6.

Forces caused by differences in chemical concentration

Only two phenomena will be discussed here, viz, diffusion and osmosis. Diffusion and osmosis are processes which take place during the movement of molecules or other particles when these are unevenly dispersed in a fluid medium.

In the case of diffusion, there is a force which causes particle movement from an area of high concentration of the particles to an area of low concentration, until equilibrium is reached. Diffusion plays a role in many functions of the human body, the most important being the supply of oxygen to the cells. Air is inhaled by the lungs and comes into contact with the blood. Because air contains a higher concentration of oxygen than the blood, the oxygen diffuses from the air to the blood. It is then transported to the tissues by the bloodvessels. Since the blood contains more oxygen than the tissues, the oxygen diffuses from the blood to the tissues.

In the case of osmosis, concentration differences exist between two areas or compartments which are separated by a membrane that allows the molecules of the solvent to pass through, but not the molecules of the solute. Movement of the solvent takes place from the area with a low concentration of the solute, to the area with a high concentration of the solute. In this way the solute tries to equalise its own molecular concentrations on both sides of the membrane. Osmosis plays a role in many functions of the human body. An example of this is the exchange of fluids (water, various nutrients, electrolytes, etc.) between the blood and the tissues. Diffusion and osmosis will be discussed in more detail in Chapter 11.

Chemotactic forces

Foreign substances or particles entering the human body often produce forces which attract the white blood cells. It has not been established whether the forces are produced by the bodies themselves or by materials which are freed when the body cells are damaged by these foreign agents. The force that attracts the white blood cells is called a chemotactic force. Because white blood cells play such an important role in the defence of the body against foreign material, such as micro-organisms, chemotactic forces must be considered as a most important defence system of the body.

Frictional forces

When an object moves on a rough surface a frictional force opposes its movement (FIG. 3.7). The frictional force exists because the surface molecules of the object and the molecules of the surface over which it moves interact with one another. The frictional force can be mathematically formulated as follows:

$$F_W = K_W M v$$

F_W = the frictional force in newton
M = mass of the object in kg
v = speed of the object in m/s
K_W = a constant that is determined by the roughness of the surfaces and therefore indicates the ease with which the two surfaces can slide over each other.

Frictional force plays a role in the flow of electricity in the tissues, in the flow of blood in the blood vessels and in the flow of gases in the respiratory passages.

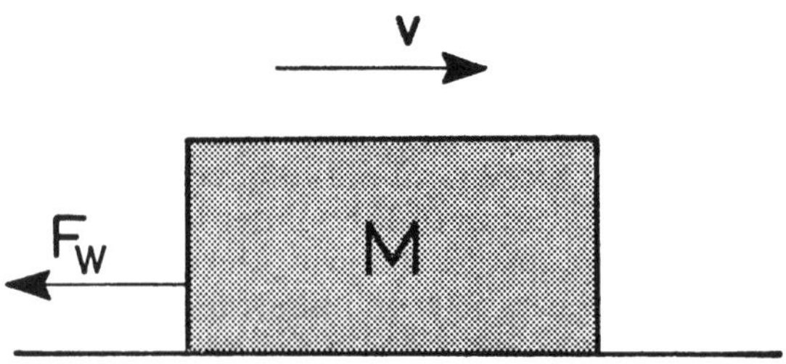

FIG. 3.7 Diagrammatic representation of a frictional force (F_W) acting on an object moving over a rough surface. M = mass of the object, v = speed of movement.

Resultant forces

In the preceding diagrams forces have been represented by arrows. The direction of the force can be represented quantitatively by the direction of the arrow and the magnitude of the force by the length of the arrow. The direction of the force is just as important as the magnitude, as is shown by the following example. When two persons are exerting an equal force in the opposite directions, no matter how great, the net force is zero. Quantities having direction as well as magnitude are called vectors. Other examples of vectors, apart from forces, are velocity and acceleration.

When two or more forces act upon the same object, the compound or resultant effect of the forces must be considered. In the case of

the forces working in the same line, there are two possibilities: 1. the forces can act in the same direction; or 2. the forces can act in the opposite directions.
In the first case the resultant force (F_R) is the sum of the two original forces (F_1 and F_2); i.e.

$$F_R = F_1 + F_2$$

In the second case the magnitude of the resultant force is equal to the difference between the two original forces. In this case the direction of the resultant force is the same as the direction of the larger of the two forces, i.e. $F_R = F_1 - F_2$.
There is of course also the possibility that the two forces are not acting in the same line (FIG. 3.8b).

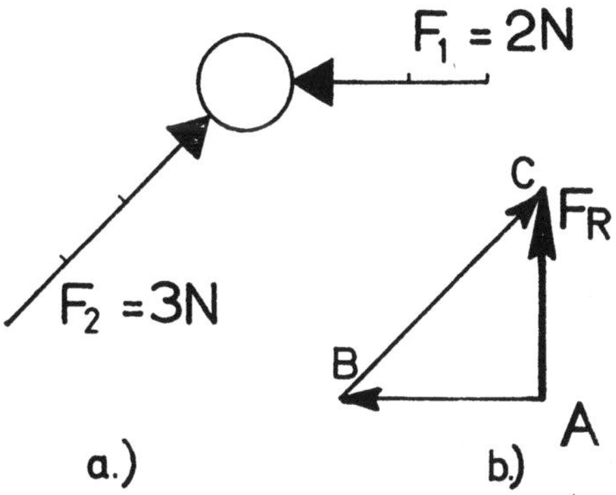

FIG. 3.8 Method of determining the resultant force of two forces acting on an object. (a) space diagram; (b) vector diagram.

To determine the resultant force the following simple graphic method is used (FIG. 3.8b).

1. Draw an arrow (AB) representing one of the forces (F_1) in magnitude and direction.
2. Draw a second arrow (BC) from the arrow head of the first in such a way that it represents the second force (F_2) in magnitude and direction.
3. Connect the rear end of the first arrow (A) with the arrow head (C) of the second. This third arrow (AC) represents the resultant force (F_R) in magnitude and direction.

This method can be extended to determine the resultant force of more than two forces (FIG. 3.9). The arrows representing the forces are drawn one after the other in such a way that the succeeding arrows begin at the arrow heads of the previous ones. The resultant force is always represented in magnitude and size by the arrow joining the rear end of the first arrow to the arrow head of the last arrow. The magnitude of the resultant force is determined by measuring the length of the arrow, and its direction is given in degrees with regard to a certain standard position.

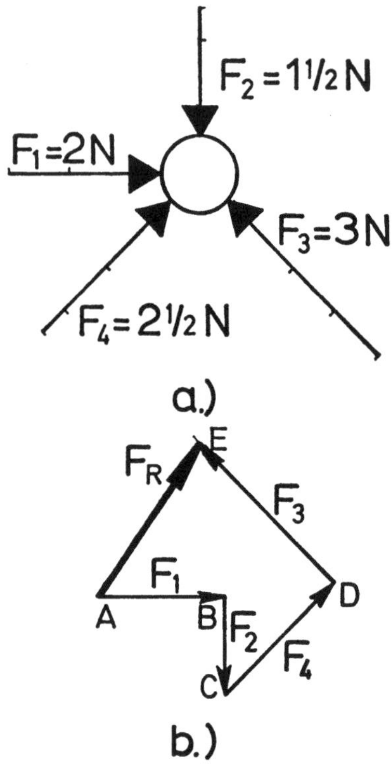

FIG. 3.9 Diagrammatic representation of the determination of the resultant force when more than two forces act on an object. (a) Space diagram; (b) vector diagram.

Work, energy and power

Work is performed when a force causes movement. Mathematically work can be described by the following equation:

$$W = F \times s$$

W = work performed in joule

F = force in newton

s = distance in metre through which the force has moved.

Example. An object (FIG. 3.10) is moved from the floor on to a table. The force used to lift the object is 10 newton and the height of the table is 0,7 m. How much work is performed?

$$W = F \times s$$
$$F = 10 \text{ newtons}$$
$$s = 0{,}7 \text{ metres}$$
$$\therefore W = 10 \times 0{,}7 = 7 \text{ joules}$$

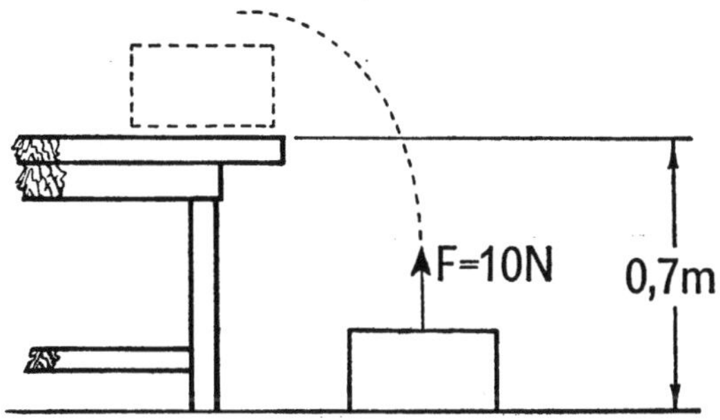

FIG. 3.10 Work is being performed when a force (F = 10N) causes movement through a distance (0,7 m).

Energy is the ability of an object to perform work or to cause movement. There are two types of energy, viz. 1. potential and 2. kinetic energy.

Potential energy is the energy inherent in a system or body by virtue of its composition or position in relation to other objects.

Example. An object is hanging by a string (FIG. 3.11). The force of the earth (5 newton) attempts to attract the object towards the earth, but the string prevents it. If the string is cut the object will move and thus perform work. The object has potential energy which is released by cutting the string. The potential energy (W_P) is equal to the work that can be performed, i.e.

$$W_P = F \times s = 5 \times 1{,}2 = 6 \text{ joules}$$

Kinetic energy is the ability of a moving object to perform work. Assume that object A (FIG. 3.12) moves with a speed, v, and collides with a standing object B. Object B is moved by object A, which means that the latter has energy, part of which has been transferred to the former. Mathematically, kinetic energy can be expressed as follows:

$$W_K = \tfrac{1}{2} Mv^2$$

W_K = kinetic energy in joule,
M = mass of the moving object in kg
v = speed of movement in m/s

Example. If the mass of the moving object is 5 kg and its speed is 2 m/s, what is its kinetic energy?

$$W_K = \tfrac{1}{2}Mv^2$$
$$\therefore W_K = \tfrac{1}{2} \times 5 \times 2^2 = 10 \text{ joules}$$

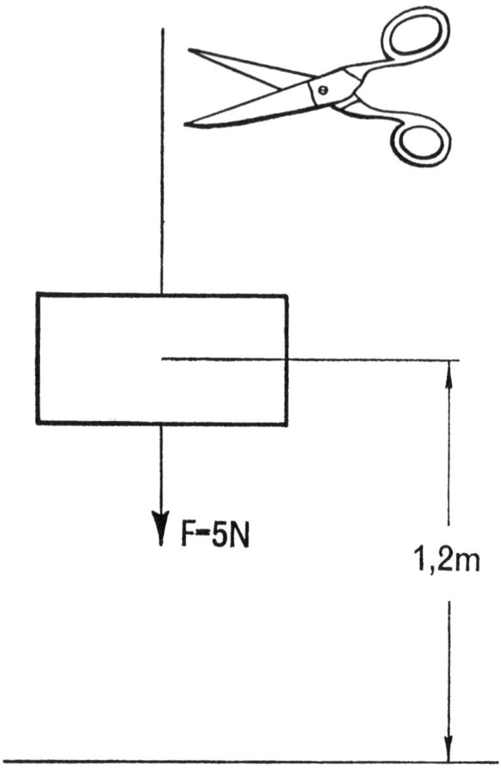

FIG. 3.11 Diagrammatic representation of potential energy. F = 5N = weight of the object. Height of the object above the ground is 1,2 m.

FIG. 3.12 Diagrammatic representation of kinetic energy. An object A, mass M, moving at a speed v can bring another object B in motion.

Power is the rate at which work is performed. In the above example (FIG. 3.10) the performed work was 7 joule. The rate of performing the work would have been different had the object been placed on the table in one second or in ten seconds. In the first case the power is 7 joule/second, and in the second case it is 7/10 = 0,7 joule/second. Mathematically, power can be expressed by the following formula:

$$P = \frac{W}{t}$$

P = power in watt (1 watt = 1 joule/second)

W = work performed in joule

t = time taken to perform the work in seconds.

SECTION II

ELECTRICITY

CHAPTER 4

The Origin of Electricity and the Flow of Electricity in Conductors

Introduction

Despite the fact that electricity is used daily in all spheres of life, and that its fundamental principles are relatively simple, many people think of it as something mysterious. *The study of electricity deals mainly with the study of the behaviour of charged particles.* The problem arises from the fact that man has no sense organ with which he can perceive charged particles. He can only sense the effects of their flow, as for example, when he puts his fingers into a wall plug, or turns on a light switch, or places the two poles of a battery on his tongue. In the first instance, his muscles will contract painfully, in the second, he will perceive light, while in the third instance he will experience an acid taste on his tongue.

Until relatively recently we were unable to see viruses, but were nevertheless aware of their existence and of their effects on tissues, including the tissues of the human body. According to the symptoms and signs of diseases produced by viruses, certain properties of these viruses could be inferred. Today we have advanced in the technical field to such an extent that we are able to see viruses by means of very strong microscopes (incidentally microscopes that are based on the principles of electricity). What is more, the viruses appear precisely as was predicted from their properties. What applied to viruses, holds for electricity today: we cannot see the structural units of electricity, even with the most sensitive microscope, but we can infer certain details about them from their behaviour.

Atoms and molecules

All substances existing in nature can be divided into *elements* and *compounds*. There are over a hundred known elements. Each of these is represented by a symbol, e.g., hydrogen by H, oxygen by O, iron by **Fe**, etc. Compounds are formed by the chemical reaction between two or more elements. For example, water is formed as a result of the combination of oxygen and hydrogen. The chemical and physical properties of the compound differ entirely from the properties of the elements constituting the compound.

Elements and compounds are built up of small particles called atoms and molecules respectively. *In the case of a compound the smallest portion that still possesses the specific properties of the compound is the molecule.* Molecules can be further broken down into atoms, in which case the original properties of the compound are lost. Elements, on the other hand, can be broken down into atoms without a loss in the original properties. In other words, *atoms are the smallest portions of elements that still possess the original properties of the element.*

Particle	*Where it is found*	*Charge*	*Mass (dalton)*
proton	nucleus	positive	1
neutron	nucleus	neutral	1
electron	orbiting around the nucleus	negative	1/1 800

TABLE 4.1 Constituents of an atom.

An atom is built up of three particles: *electrons, protons and neutrons.* Protons and neutrons together form the nucleus of the atom (FIG. 4.1 and TABLE 4.1). The mass of protons and neutrons is 1 800 times that of the mass of electrons.* Protons are positively charged, electrons are negatively charged and neutrons are electrically neutral. The magnitude of the electrical charges carried by protons and electrons is equal. A stable atom possesses the same number of protons and electrons and is therefore neutral.

Elements differ from each other in that their atoms do not possess the same number of protons, and they can be classified according to the number of protons in their atoms. For example, hydrogen is the first element because its atom contains one proton; helium is the second, with two protons, and oxygen the eighth. The total number of protons in an element is referred to as the *atomic number.*

The mass of an atom is determined by the total number of protons, neutrons and electrons it contains. The mass of electrons is relatively small and their effect on the atomic mass is negligible. As already mentioned, the mass of the proton is more or less equal to the mass of the neutron, therefore the mass of protons and neutrons is used as the unit of *atomic mass.* The unit of atomic mass is the dalton. (Atomic mass (in dalton) = number of protons + number of neutrons.) The hydrogen atom consists of a proton and electron only, and its atomic mass is therefore one dalton. The helium atom consists of two protons, two neutrons and two electrons and has an atomic mass of

*Absolute mass of a proton is 1,672 $\times$ 10^{-27} kg and of an electron 0,910 9 $\times$ 10^{-30} kg.

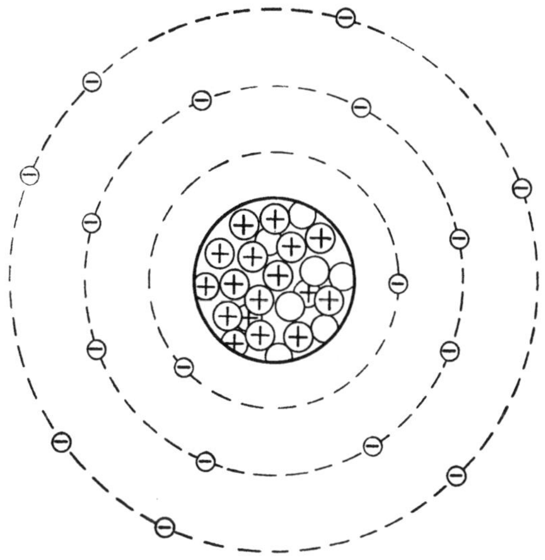

FIG. 4.1 Schematic representation of the structure of an atom. ⊖ = electron. ⊕ = proton, O = neutron

four dalton. The symbol, the atomic number and the atomic mass of an element is generally given as follows:

$$_{Z}X^{A}$$

X = symbol of atom

Z = number of protons = number of electrons = atomic number

A = number of protons + number of neutrons = atomic mass.

According to this notation hydrogen is given as $_{1}H^{1}$, oxygen as $_{8}O^{16}$, nitrogen as $_{7}N^{14}$.

Ions

In the event of an atom losing or gaining an electron, the atom is said to become ionised. This means that the atom no longer possesses the same number of protons and electrons. It therefore carries a positive or negative charge and is called a positive or negative ion respectively. A particular atom is always ionised in such a way as to give rise to a particular ion. For example, hydrogen loses an electron to become a positive ion, which is then shown as, $_1H^{1+}$ or simply H^+. When chlorine ionises, it takes up an electron, and therefore becomes a negative ion, namely $_{17}Cl^{35-}$ or simply Cl^-.

Isotopes

One of the most important attributes of an element is the number of protons in its nucleus. This is important because the number of protons, which always equals the number of electrons, determines the chemical behaviour of the element. The number of neutrons in an element can vary without the chemical characteristics of the element changing. If, for example, the number of protons in hydrogen remains unchanged, but the number of neutrons changes, we still have the same element with the *same chemical properties*, despite the fact that the atomic mass has changed. This means that hydrogen keeps its chemical properties whether it has no neutrons or whether it has one or two neutrons in its atom ($_1H^1$, $_1H^2$, $_1H^3$). These various forms of the element are called the isotopes of the element.

Free electrons in a metal

Electrons move in orbits around the nucleus of an atom and the *total number of electrons in a fixed orbit is constant.* The orbit closest to the nucleus never contains more than two electrons, the next never more than eight, the next never more than eighteen, and so on. This means that in an element containing many electrons the electrons in the outside orbit are farthest removed from the nucleus. As has been explained (Chapter 3), the forces of attraction on an electron diminish rapidly the farther away the electron is from the nucleus. This means that *these electrons can free themselves easily from the force of attraction of the nucleus if external forces act upon them.* If an electron frees itself it becomes a *free electron* and the atom becomes a positive ion. It is these free electrons that make the flow of electricity possible.

Copper possesses many free electrons, and is therefore a *good conductor* of electricity. Materials with few free electrons are bad conductors and are called *insulators.* There is also an intermediary group of materials called *semi-conductors.* These play an important role in electronics. Metals are good conductors of electricity, but good conductors also vary in their number of free electrons. For example, copper has more free electrons than aluminium and is therefore a better conductor of electricity. The ability to conduct electricity is therefore determined by the material itself.

'Flow' of electrons

For our purposes it is sufficient to define electrical current as the movement of charged particles. In order to bring about movement, *force* is necessary. If, therefore, an electrical force is created in the metal, the free electrons will move in an orderly fashion producing a current. The greater the number of electrons that move together within a certain unit of time, the stronger is the *electrical current* that flows.

When one end of a metallic wire is positive with respect to the other end, electrons will flow from the negative to the positive end (FIG. 4.2). The question now arises as to how fast the electrons flow through such a wire? When the switch of a lamp is put on, the electric lamp glows almost immediately, *yet the flow of electrons is very slow*. A conductor of electricity contains millions of free electrons. It is actually packed with electrons to such an extent that there is no place for new electrons to enter. When an electron is forced in at one point, another must fall out at some other point, while all the other electrons move along slowly. The wire can be compared with a bus full of people. If someone forces himself in at the back door, another will fall out at the front door. This will happen within seconds, yet the people in the bus itself move slowly forward. The movement of electrons in a metal such as tungsten gives rise to friction and thus produces light and heat. Electrons move at a constant rate. If this were not so, the rate would increase all the time, and after a while the electrons would move at a tremendous speed. Should, however, the rate of movement decrease continuously, the electrons will come to a halt after a while and the flow of current will cease.

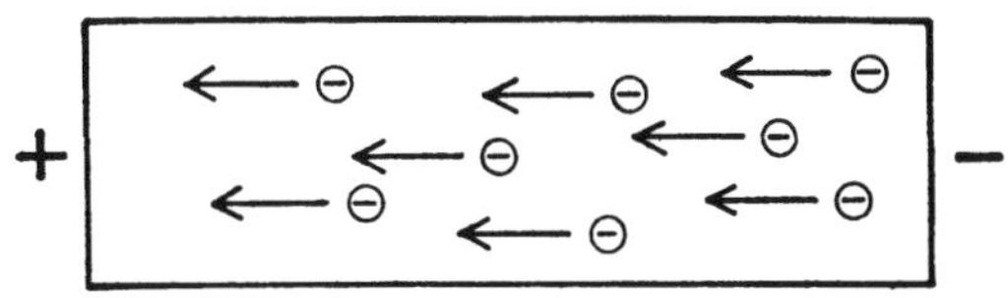

4.2 Electrons flow from the negative to the positive terminal of a conductor.

Electromotive force

From the previous discussion it is evident that electrons can only perform useful work if they move together. To make them move,

an *electrical force* is necessary. If one end of the metal is made positive with respect to the other end, the electrons will move from the negative to the positive end. The simplest way to make a piece of metal negative is to add electrons. To make it positive, one takes away electrons. "*Positive*" *thus means a relative shortage of electrons and* "*negative*" *a relative surplus of electrons.* A metal that is positive or negative in relation to the original state is called a *charged* metal.
In nature there is a constant tendency towards equilibrium. Water will for instance flow until the level is the same at all points in a system. In a similar way chemical reactions proceed until equilibrium is reached. The same holds for electricity. If there is a difference in the electron concentration between two points in a piece of metal, the electrons will flow until their distribution is the same throughout the metal.
A difference in charge causes electrons to flow, and the force causing the flow is called the *electromotive force* (*EMF*), *potential difference* or *tension.* The unit of EMF is the *volt* (V). A cell (e.g. a torch battery) is the best-known source of constant EMF. The principles of a cell are discussed in Chapter 5.

Opposing electrical force

In Chapter 3 reference was made briefly to two of Newton's laws. In order to understand the development of an opposing force, one must know the second law of Newton (N_2). According to this law, an object *accelerates proportionally to the force acting upon it* (FIG. 4.3). This means:

$$a = kF$$

F = force in newton

a = acceleration in m/s^2

k = a constant

This means that a constant force causes a constant acceleration. *Acceleration* means an increase of speed per unit of time. The difference between speed and acceleration can be illustrated by the following example. Assume that a standing vehicle begins to move, and after the first second its speed is 0,1 km/s, after the second 0,2 km/s and after the third 0,3 km/s. After each second the vehicle moves

0,1 km/s faster. The acceleration is therefore constant, i.e. 0,1 km/s/s or 0,1 km/s². The *speed* on the other hand, is the distance travelled per unit of time and it varies as acceleration takes place.

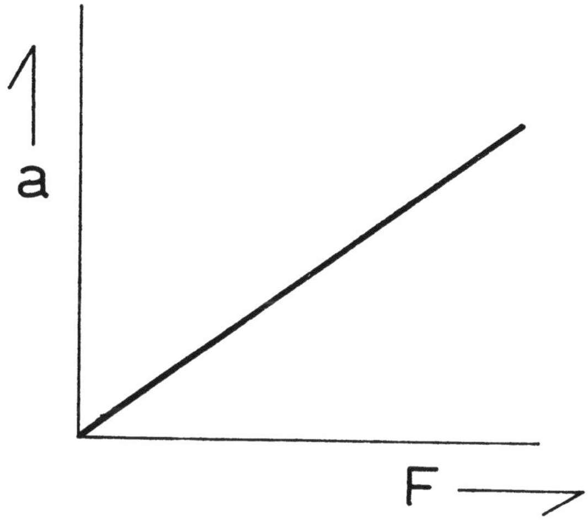

FIG. 4.3 Graphical representation of the relationship between a force (F) acting on an object and the acceleration (*a*) of the object.

Frictional force (see again, Chapter 3) is produced when an object moves over a rough surface, the friction opposing the motion. The frictional force that is produced is, within limits, proportional to the speed against which the object moves (FIG. 4.4), or in brief

$$F_r = K_r v$$

F_r = frictional force in newton

v = speed in m/s

K_r = a constant

When an electron moves in a conductor it is also subject to an opposing force caused by "friction" between the electron and the structure of the conductor.

With the help of the above two equations, the movement of an electron under the influence of the electromotive force can be examined. According to N_2, the electromotive force (F_E) causes acceleration of the electron, i.e. it will move faster and faster. As a result of the increasing speed (v) the opposing force (F_r) increases and therefore the resultant force exerted on the electron diminishes (resultant force = $F_E - F_r$). When the electron reaches a certain speed, the two forces become equal. This means that the resultant

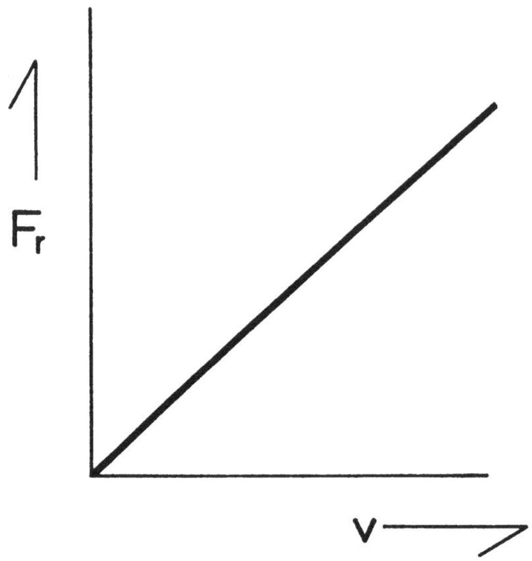

FIG. 4.4 Graphical representation of an opposing force (F_r) acting on an object moving at a speed = v

force is zero and the electron will not accelerate, but will *continue at a constant rate* until external forces influence its motion (N_1). Mathematically this can be expressed as follows:

$$F_E = F_r = K_r v$$

The unit of the electromotive force and of the opposing force is the *volt* (V) and it is customary to use the symbols E and V for F_E and F_r respectively. This means

$$E = V = K_r v$$

Electrical current intensity and resistance

The principles according to which an electron moves apply also to electrons when they are flowing together. The electrical current intensity is defined as the number of *electrons* moving past a certain point in a wire *per unit of time*. The faster the electrons move, the stronger will be the current, i.e., $v = k_1 I$

where I $=$ current intensity

v $=$ mean speed of electrons

k_1 $=$ a constant

Substituting for $v = K_1I$ in the equation $V = K_R v$

$$V = K_r(k_1I) = (K_rk_1)I = RI$$

This means that when an electrical current flows in a wire the current intensity and the potential difference or tension are proportional to each other. The constant K_rk_1 is called the resistance of the conductor and is shown by the symbol R. This relation between the tension, current intensity and resistance is called *Ohm's Law.* The equation $V = IR$ can also be written as $I = V/R$. This means that for a constant electromotive force the current is inversely proportional to the resistance, i.e., the current increases as the resistance decreases. The resistance can therefore limit the current in a conductor. The unit of current is the *ampere* (A) and that of resistance, the *ohm* (Ω).

Problem 4.1. Calculate the current intensity in a conductor having a resistance of 10 ohms, when an electromotive force of 10 volts is connected between its two ends.

$$E = V = 10 \text{ volts}$$
$$R = 10 \text{ ohms}$$
$$\therefore \quad I = \frac{V}{R} = \frac{10}{10} = 1 \text{ ampere}$$

Resistance is a property of the conductor and depends on the number of free electrons per unit volume of the material of the conductor as well as on the dimensions of the conductor. It can be calculated with the aid of the following formula:

$$R = \rho \frac{L}{A}$$

R $=$ resistance in ohm

ρ $=$ resistivity of the material at a certain temperature in ohm-m

L $=$ length of the conductor in m

A $=$ cross-sectional area of the conductor in m^2

Specific resistance, or *resistivity* is a property of the material from which the conductor is manufactured. It is determined both by the number of free electrons per unit volume at a certain temperature and by the dimensions of the conductor. The resistivity of copper at °C for example is $1{,}76 \times 10^{-8}$ ohm—m.

Problem 4.2. Calculate the resistance of a copper wire, having a length of 1 km and a cross-sectional area of $1 \times 10^{-8}\text{m}^2$

$$\text{Use the formula } R = \rho \frac{L}{A}$$

$$\rho = 1{,}76 \times 10^{-8} \text{ ohm—m}$$

$$L = 10^3\text{m}$$

$$A = 10^{-8}\text{m}^2$$

$$\therefore\ R = \frac{1{,}76 \times 10^{-8} \times 10^3}{10^{-8}} = 1{,}76 \times 10^3 \text{ ohms}$$

It is therefore possible to construct a conductor with a certain predetermined resistance by choosing the appropriate material and dimensions. It is customary to speak simply of a *resistor* when a conductor with a certain resistance is implied. Resistors with all possible values are available at a very low price.
The resistivity of metals varies as the temperature changes. An interesting application of this principle is to be found in the determination of the respiratory rate of a patient. A wire in which a current is flowing is held in front of the nose of the patient. The current that flows in the wire produces heat. When the patient breathes in through his nose, the air moving over the wire carries away the heat. The periodic cooling of the wire that takes place is therefore determined by the respiration of the patient. The change in temperature causes a change in resistance and therefore a change in the current. This variation in current can be registered and the rate of variation gives an indication of the respiratory rate of the patient.

Resultant resistance

When several resistors are connected in an electrical circuit their resultant resistance can be calculated. Resistors can be connected *in parallel* or *in series* (Fig. 4.5).

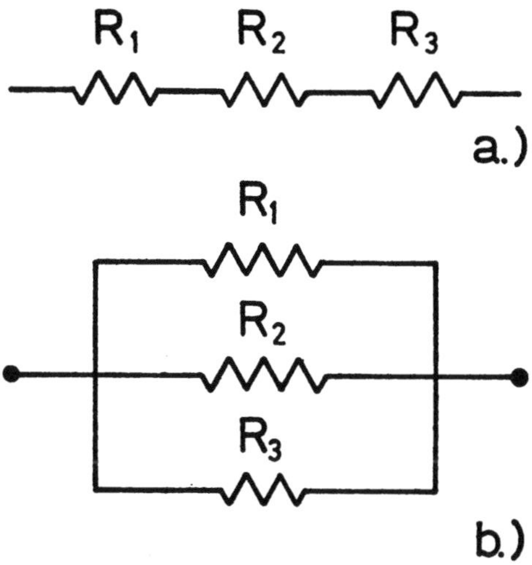

Fig. 4.5 Connection of resistances: (*a*) in series, (*b*) in parallel. The symbol -/\/\/\- is used to indicate a resistance.

When resistors are connected in series, the resultant or total resistance is simply the sum of all the individual resistances, i.e.,

$$R_{TOT} = R_1 + R_2 + R_3 + \ldots\ldots R_n$$

To calculate the resultant resistance of resistors in parallel, the sum of the reciprocals of the values of the individual resistances must be obtained. The answer obtained is the reciprocal value of the resultant resistance. Algebraically this can be written

$$\frac{1}{R_{TOT}} = \frac{1}{R_1} + \frac{1}{R_2} + \frac{1}{R_3} + \ldots\ldots \frac{1}{R_n}$$

The calculation of the resultant resistance can best be illustrated by the following problems.

Problem 4.3. Calculate the resultant resistance if the following resistors are connected in series: 5 ohms, 8 ohms and 3 ohms.

$$R_{TOT} = 3 + 8 + 5 = 16 \text{ ohm}$$

Problem 4.4. Calculate the resultant resistance if the resistors in in Problem 4.3 are connected in parallel

$$\frac{1}{R_{TOT}} = \frac{1}{3} + \frac{1}{5} + \frac{1}{8} = \frac{79}{120}$$

$$\therefore \quad R_{TOT} = \frac{120}{79} = 1{,}5 \text{ ohms}$$

N.B. — When the resistors are connected in parallel the total resistance is always smaller than the smallest of the individual resistances.

Problem 4.5. In FIG. 4.6 a mixed connection is shown. Calculate the total resistance between the points A and B.

$$\frac{1}{R_{DE}} = \frac{1}{12} + \frac{1}{6} = \frac{3}{12}$$

$$\therefore \quad R_{DE} = \frac{12}{3} = 4$$

$$R_{CE} = 4 + 8 = 12$$

$$\frac{1}{R_{CF}} = \frac{1}{12} + \frac{1}{6} = \frac{3}{12}$$

$$\therefore \quad R_{CF} = \frac{12}{3} = 4$$

$$R_{AB} = 4 + 4 = 8 \text{ ohm}$$

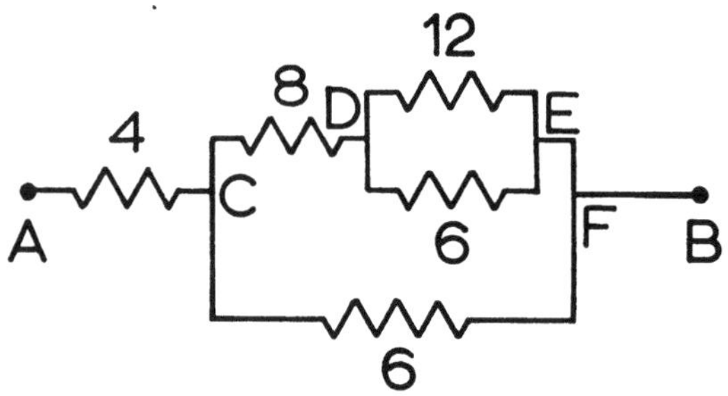

FIG. 4.6 Schematic representation of problem 4.5.

Problem 4.6. An electric lamp, a resistor and a battery are connected in series (FIG. 4.7). The battery, shown by symbol —|⊢— produces a 10 volt electromotive force. The lamp, shown by symbol ⊕, has a resistance of 100 ohms and produces light when a current of 0,01 ampere flows through it. What must the resistance of the resistor be in order to obtain the correct current intensity?

$$I = ,01 \text{ ampere}$$

$$V = 10 \text{ volts}$$

$$R_{TOT} = \frac{V}{I} = \frac{10}{0,01} = 1\,000 \text{ ohms}$$

The resistor is connected in series with the lamp.

$$\therefore \quad R_{TOT} = R + 100 = 1\,000 \text{ ohms}$$

$$\therefore \quad R = 1\,000 - 100 = 900 \text{ ohms}$$

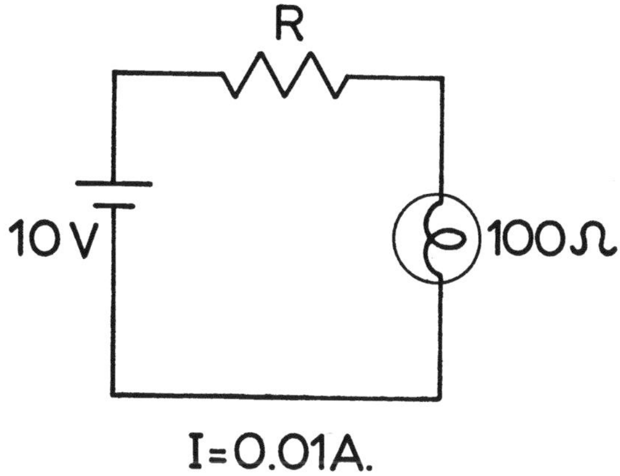

FIG. 4.7 Schematic representation of an electrical circuit consisting of a 10 V battery (symbol ⊥ 10 V), a lamp (symbol ⊗) having a resistance of 100 Ω and an unknown resistance R. The current is 0,01 A.

The flow of electricity through liquids

In metals the conducting properties are determined by the concentration of free electrons and in *liquids*, by the concentration of *ions*. Water consists of water molecules of which a certain number are ionised. The reaction is as follows:

$$H_2O \rightleftharpoons H^+ + OH^-$$

This means that H^+-ions and OH^--ions are produced when water molecules ionise (dissociate). The arrows pointing in opposite directions indicate a state of equilibrium. This means that if too little water is ionised, it proceeds from right to left. The equilibrium point is determined by the following equation:

$$[H^+] \times [OH^-] = K[H_2O]$$

where $[H^+]$, $[OH^-]$ and $[H_2O]$ respectively indicate the concentrations of H^+, OH^- and H_2O.
According to this relationship, equilibrium is reached when the product of the H^+ and OH^- concentrations equals a constant fraction multiplied by the concentration of undissociated water. This means that only a *certain fraction of water dissociates*, and since this

fraction is small there are therefore only a few ions to propagate the flow of electricity.

If ordinary table salt (NaCl) is dissolved in water, it dissociates as follows:

$$NaCl \rightleftharpoons Na^+ + Cl^-$$

As in the case of pure water, only a fraction of the NaCl ionises before equilibrium is reached, but the fraction in this instance is much higher than in the case of water and therefore a solution of NaCl is a much better conductor of electricity than is pure water.

Movement of ions in liquids

An electromotive force is necessary to make ions move in a certain direction. The electromotive force (tension or potential difference) can be created in the liquid by producing a relative electron surplus at one point and a relative electron deficiency at another. This can be done with the aid of a battery, and two metal plates (electrodes) connected to the two poles of the battery. The electrodes are then immersed in the liquid (FIG. 4.8). The electrode connected to the positive pole of the battery is called the anode, and the other connected to the negative pole is called the cathode. Negative ions move towards the anode and are called anions; positive ions move towards the cathode and are called cations. When an anion arrives at the anode it yields up its excess electrons because the anode is deficient in electrons. The function of the battery is to maintain the charge difference and it therefore causes the movement of the extra electron *from the anode to the cathode.* When a cation arrives at the cathode it receives an electron from the cathode. To keep the charge difference constant, the battery must again bring about the movement of an electron *from the anode to the cathode.* The movement of both positive and negative ions brings about the same type of movement as is the case of electrons in a wire, i.e. from anode to cathode, and consequently the total effect of ionic movement is equal to the individual ionic movements. This can be described briefly as follows:

$$I = I_{Na^+} + I_{Cl^-}$$

I = total current

I_{Na^+} = current caused by the Na^+ ions

I_{Cl^-} = current caused by the Cl^- ions.

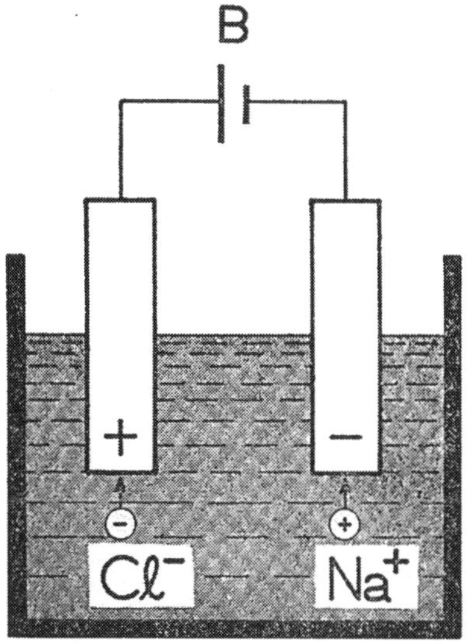

FIG. 4.8 Schematic representation of the movement of ions in a liquid. A battery (B) produces positivity (+) and negativity (—) on the two metal plates, respectively. Negative ions move to the positive electrode and positive ions to the negative electrode.

In metallic conductors such as a conducting wire the electrons move in straight lines along its longitudinal axis. Liquid conductors are not long wires, but three dimensional volumes and ions are therefore able to move in all directions in the whole volume. This type of conductor is thus called a *volume conductor.*

FIG. 4.9 portrays a section of a volume conductor with a positive electrode in its centre. The volume conductor contains a salt solution, i.e., Na^+ and Cl^- ions are freely available in the solution to conduct electricity. In the figure only a single Na^+ ion is shown. The positive electrode (anode) exerts a repelling force on the ion and forces it away. The ion moves in a straight line away from the electrode.

FIG. 4.10 represents a volume conductor with a positive and a negative electrode (this pair is often called a dipole) and a single Na^+ ion that is initially in position 1. Two forces act upon the ion, namely the repelling force due to the anode, and the force of attraction due to the cathode. In position 1 the former is greater because the ion is closer to the anode. The resultant force F^1 can be determined by the method described in Chapter 3. The ion moves in the

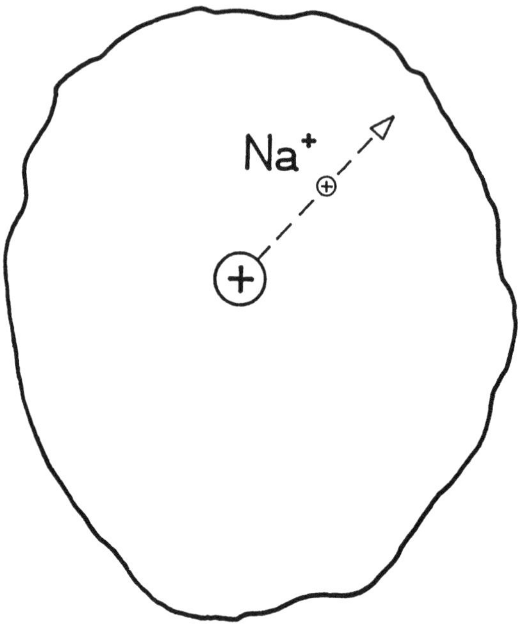

FIG. 4.9 Schematic representation of the influence of a positive electrode (anode) on a positive ion in a liquid.

direction of the resultant force. Assume that after a certain time it arrives at position 2. The two forces on the ion are now different from those active in position 1, and consequently the ion moves in another direction. This means that the electrons do not simply move in a straight line between the electrodes, but follow a *path* that is determined by the changing resultant force.

Equipotential lines and surfaces

The magnitude of the force exerted upon the ion represented in FIG. 4.9 changes as the distance of the ion from the electrode changes, yet remains the same at various points all around but equidistant from the electrode. Since the other name for electrical force is potential difference, the line through these points at which the force

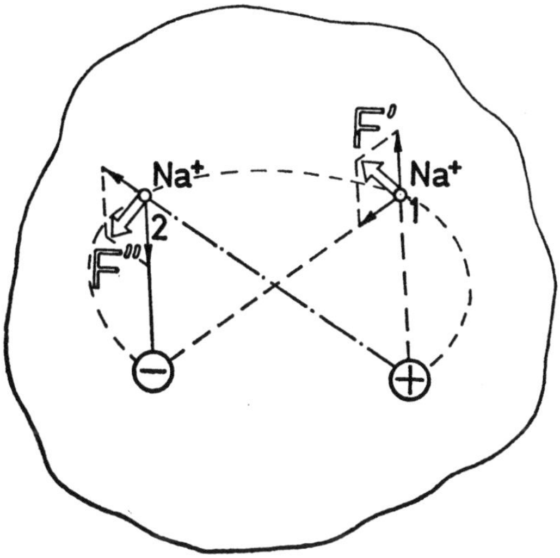

FIG. 4.10 Schematic representation of the joint effect of a positive and a negative electrode on a positive ion in a volume conductor. The numbers 1 and 2 indicate two positions of the ion. F^{1} and F^{11} are, respectively, the total forces at the two positions.

is the same is called the *equipotential line*. It must be remembered that a volume conductor is three dimensional and one should rather speak of *equipotential surfaces*. The whole system of equipotential lines or surfaces is called the *electrical field*.

Similar circular electrical fields develop around the two electrodes in FIG. 4.10. It can be seen (FIG. 4.12) that the two electrical fields influence each other in such a way that the lines between the electrodes move closer to one another while the lines outside the electrodes spread out. The lines spread out to the surfaces of the volume conductor and cross it at various points. There exists, therefore, a potential difference between various points on the surface. This potential difference can be measured with special instruments. The unit of potential difference is the volt and the instrument with which it can be measured is called a *voltmeter*.

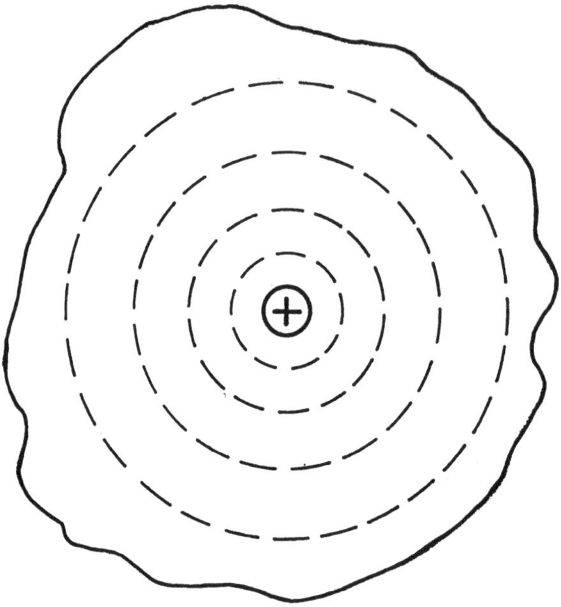

FIG. 4.11 The electric field around a positive electrode. The concentric circles are called the equipotential lines because the electrical force which acts upon all the ions present on a specific line is the same.

The principles of conduction of an electrical current through volume conductors have their applications in physiology. The human body is a volume conductor containing a salt solution; the skin acts as a container. As it will be shown later, various tissues of the body, including nerves, muscles and glands, transmit electrical dipoles. The electrical field produced by these dipoles spreads out to the skin. This means that potential differences exist between various points on the surface of the body. These potential differences can be measured by means of two electrodes placed on the skin. This principle is used in the construction and operation of such diagnostic instruments as the electrocardiograph, the electromyograph and the

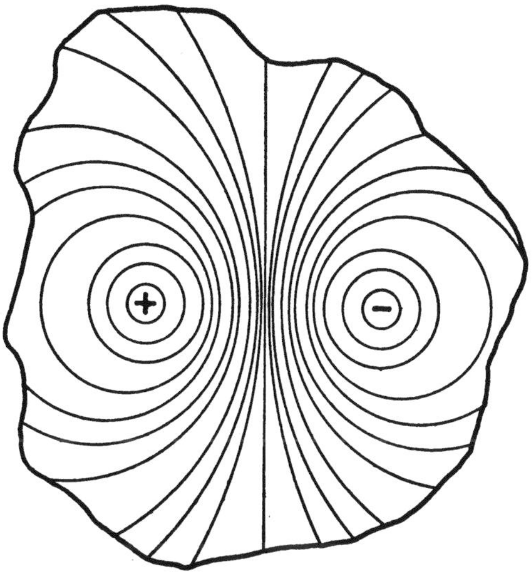

Fig. 4.12 The distribution of the equipotential lines around a dipole.

electroencephalograph which measure the electric activity of the heart, skeletal muscle and the brain respectively. The diagnostic use of these instruments is based on the fact that lesions in the organs mentioned, disturb the normal production and distribution of body electricity.

CHAPTER 5

Production of Electricity in Electrical and Living Cells

Introduction

Electromotive force is necessary to cause movement of charged particles. The simplest way of producing an electromotive force is by means of an *electrical cell*, for example a torch battery, which *changes chemical energy into electrical energy*. An electrical cell works according to the principle that different chemical reactions proceed at various rates, i.e., some occur faster than others.
In the human body, *living cells* produce their own electricity by changing metabolic energy into electrical energy. This electricity is used for activities of the body such as contraction of muscle and impulse transmission in nerve.

Production of electricity in an electrical cell

The oldest and simplest electrical cell was constructed by Volta 200 years ago. The Voltaic cell uses a solution of sulphuric acid with two metallic electrodes, one of zinc and the other of copper, immersed in the acid (FIG. 5.1).

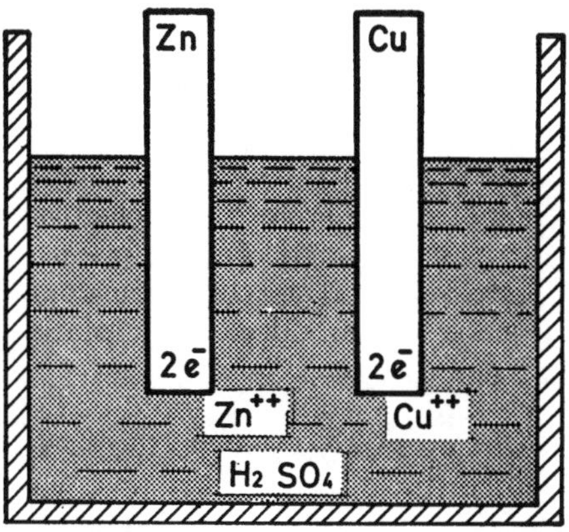

FIG. 5.1 Schematic representation of the principles of a Voltaic cell (see text).

As it has been mentioned (Chapter 4), the sulphuric acid dissociates to yield hydrogen ions (H^+) and sulphate ions ($SO_4^=$):

$$H_2SO_4 \rightleftarrows 2H^+ + SO_4^=$$

Equilibrium is reached between the ionised and non-ionised parts of the sulphuric acid when

$$[H^+] \times [SO_4^=] = K \times [H_2SO_4]$$

The zinc electrode dissolves slowly at the point of contact between the electrode and the acid; the Zn^{++} is released but the electrons stay on the metal, i.e.

$$Zn \rightarrow Zn^{++} + 2e^-$$

As a result of this reaction electrons collect on the electrode. The zinc ions released by the electrode react with the sulphate ions as follows

$$Zn^{++} + SO_4^= \rightleftarrows ZnSO_4$$

At the copper electrode similar reactions take place, viz,

$$Cu \rightarrow Cu^{++} + 2e^-$$
$$\text{and } Cu^{++} + SO_4^= \rightleftarrows CuSO_4$$

Electrons also collect on the copper electrode. The number of electrons present on the electrodes after a certain time is determined by the *rate of the reactions at the electrode in question.* The reaction at the copper electrode proceeds at a slower rate than at the zinc electrode, and consequently the electron charges on each electrode differ. The zinc electrode acquires more electrons than the copper electrode and will therefore be negative with respect to the copper electrode. This in turn is therefore positive with respect to the zinc electrode. It is important to note that absolutely positive or negative electrodes are unnecessary. It is necessary, however, that their charges differ.

The charge difference and therefore the electromotive force produced in a cell is a *property of the chemical reactions* that take place at the plates (electrodes) and it therefore remains constant. If electrons are removed from or added to the plates the *chemical reactions will proceed to ensure that the electromotive force remains constant.* The Voltaic cell produces an electromotive force of 1,06 volt. There are other cells which function more or less on the same principles as the Voltaic cell, but which make use of other chemicals and metals. In all of them the EMF is determined by the chemical reactions. In Table 5.1 the different types of cells with their electromotive forces are shown.

Type of cell	*Electromotive force (volt)*
Voltaic	1,06
Daniell	1,09
Weston	1,18
Clark	1,43
Edison	1,5
Leclanché	1,5
Mercury	1,42
Nickel-cadmium	1,22

TABLE 5.1 The electromotive force produced by different cells.

When two or more cells are connected in series they form a battery. The EMF of a battery is the sum of the EMF of the individual cells. For example a battery that consists of 6 cells each having an EMF of 1,5 volt, has an EMF of $6 \times 1,5 = 9$ volts. Usually the cells of a battery are enclosed in a container and the whole battery is provided with two terminals. The symbols used to indicate cells and batteries are shown in FIG. 5.2.

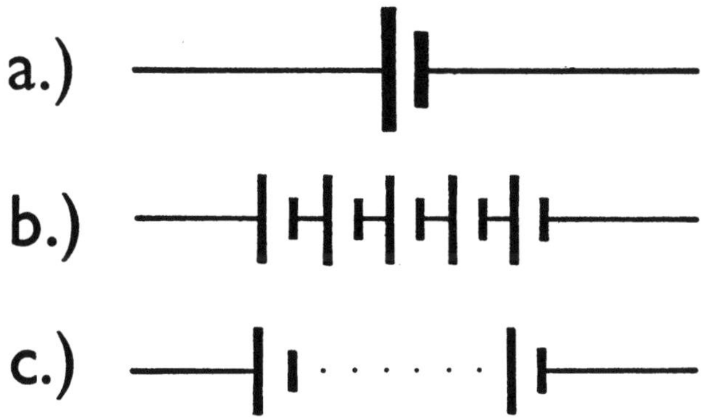

FIG. 5.2 Schematic representation of (*a*) a single cell — the longer vertical line indicates the positive pole of the cell; (*b*) and (*c*) cells connected in series to form a battery.

A battery with cells which can be recharged after they have stopped producing electromotive force, is called an accumulator. A good example of an accumulator is the battery of a car. During the charging process, a current is sent through the battery in the opposite direction to normal flow. This causes the regeneration of the original chemical compounds and the battery can be used again.
The amount of electricity that can be produced by a cell (*capacity*) is determined by the amount of chemicals available in the cell and

is thus constant for a given cell. When current flows between the electrodes of a battery, the charge difference tends to disappear, but chemical reactions proceed which keep the EMF constant. The available chemicals are therefore used, and the greater the current, the greater the ion consumption will be and the shorter the life of the cell. Mathematically this can be written as follows

$$I = \frac{K}{t}$$

where

I = current in amperes

t = time for which current flows in hours

K = capacity of the cell in *ampere-hours* (Ah)

A cell with a capacity of 1 ampere-hour (Ah) will last for 1 hour if it delivers 1 ampere. If the current drops to half this intensity, the time doubles; if it drops to a tenth of the intensity, the time becomes 10 times longer. It is possible to buy cells with different capacities. The greater the capacity, the larger the volume of the cell must be to contain the necessary chemicals. It is important to note that cells degenerate even when not in use. The maximum period a cell can be kept on the shelf without being used is called the *shelf-life* of the cell.

Problem 5.1. A battery, EMF 6 V and capacity 0,5 Ah, is used to produce current for a certain apparatus. How long will the battery last if the resistance of the apparatus is 600 ohms (FIG. 5.3)?

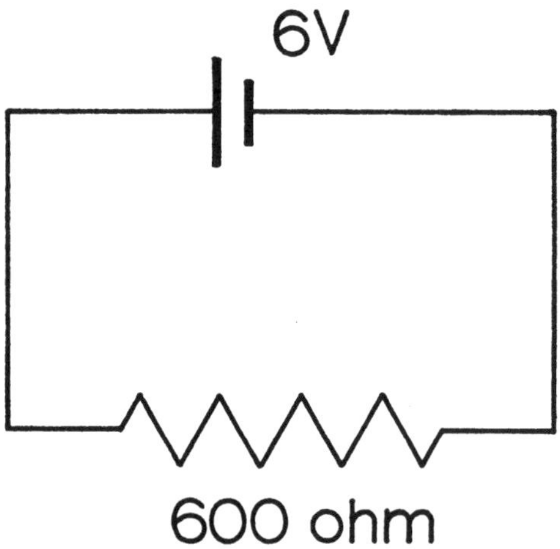

FIG. 5.3 Schematic representation of the electrical circuit outlined in problem 5.1: a 6 volt battery is connected in series with a resistance of 600 ohms.

The current that flows is

$$I = \frac{6}{600} = 0{,}01 \text{ A}$$

The time for which this current can be produced

$$t = \frac{K}{I}$$

$$= \frac{0{,}5}{0{,}01} = 50 \text{ hours}$$

Problem 5.2. In patients with a disease called total heart block, the rhythm of the heart can be kept normal by artificial means. The apparatus that is used for this purpose is called an artificial pacemaker, and is usually implanted in the body of the patient. The heart needs a mean current of 0,01 mA to render the pacemaker effective. Because a pacemaker must obviously be small, the largest battery that can be used is a 6V, 480 mAh mercury battery. The problem is:

(i) how many cells must be used in the battery in order to produce 6V?

(ii) after how many years must the pacemaker be replaced?

Answer

(i) A mercury cell produces 1,42 V and therefore the number of cells that must be used is $\frac{6}{1{,}42} \simeq 5$

(ii) $t = \frac{K}{I}$

$$t = \frac{480 \times 10^{-3}}{0{,}01 \times 10^{-3}} = 48\,000 \text{ hours}$$

$$= 2\,000 \text{ days}$$

$$= 5 \text{ years}$$

Production of electricity by the living cells of the body—resting membrane potential

In living cells a potential difference exists between the outside and the inside of the cells, i.e. across the cell membrane. When the cell is in a resting state the *ou side of the cell membrane is positive with respect to the inside of the membrane.* This potential difference, which is the result of a charge difference between the outside surface and the inside surface of the membrane, is called the resting membrane potential. It is produced as a result of 1. the cells being surrounded by a semipermeable membrane and 2. the difference in ion concentrations inside and outside the cell.

The relative concentrations of the most important ions inside and outside the cells are shown in FIG. 5.4. The fluid inside the cell is called the intracellular fluid (I) and that outside the cell, the extracellular fluid (E). The extracellular fluid contains mainly sodium ions (Na^+) and chlorine ions (Cl^-), while the intracellular fluid contains mainly potassium ions (K^+) and negatively charged protein ions (A^-).

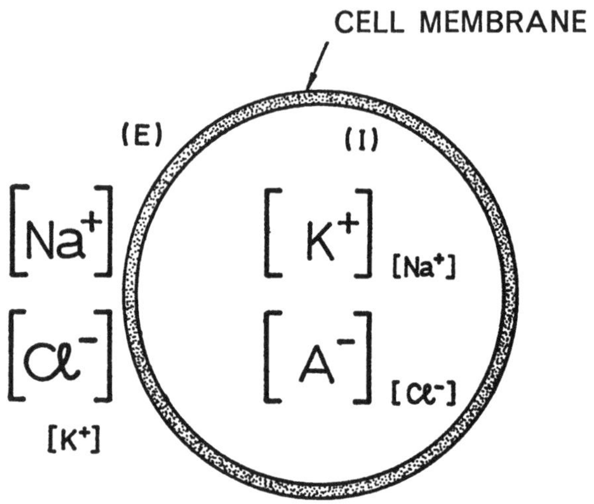

FIG. 5.4 Schematic representation of the relative ionic concentrations in the extracellular (E) and intracellular (I) fluid of the body. The Na+ and Cl− are present in large concentrations outside the cells, while K+ and the negatively charged proteins (A−) are present in large concentrations inside the cells. [X] = concentration of the substance X.

The cell membrane has small openings or pores, which allow certain substances to pass through, but prevent the passage of others. If a substance passes through with ease, the membrane is said to have a *high permeability* for that substance. If the permeability of the membrane for substance A is zero it means that that substance cannot pass through the membrane at all. Various degrees of permeability exist, the effective permeability depends on the size, shape and electrical charge of the ions concerned.*

*In water dissolved ions combine with water ions to form ion complexes. The size of the ion complex is dependent on the size of the original ion and its affinity for water ions. Of the ions important to this discussion, Na+ ion complexes are larger than the K+ and Cl− ion complexes. Although we shall use the term 'ion' it must be remembered that these are ion complexes.

In Fig. 5.5 the relative sizes of ions are illustrated schematically. The pores are lined with Ca^{++} ions. The effective size of the pores is such that K^+ and Cl^- can pass through with ease while Na^+ pass only with great difficulty and A^- not at all.

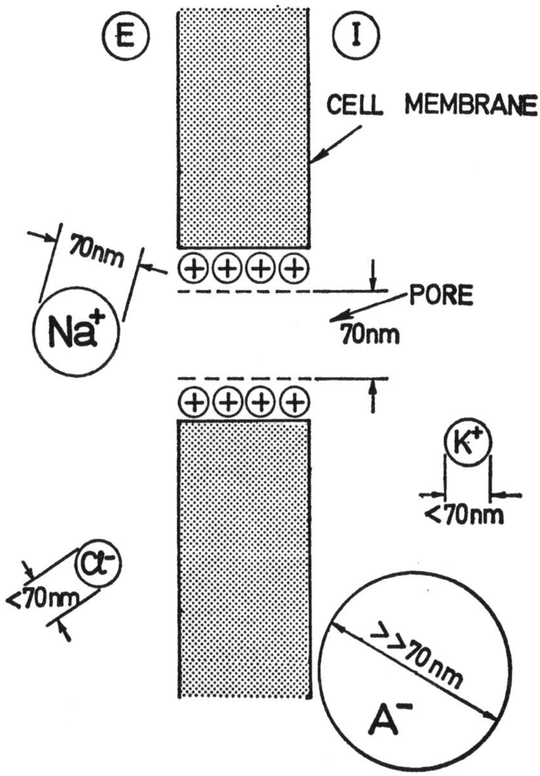

FIG. 5.5 The relative diameter of some ion complexes and pores. The pores are lined with Ca++. The effective diameter of the pores is 70 nm. Ions with an effective diameter of less than 70 nm pass through freely; those with a larger diameter do not pass through at all; those having the same diameter as the pores pass through with difficulty.

Fig. 5.6 illustrates the way in which the resting membrane potential arises. In the figure only a portion of the membrane is shown with a single pore. The concentration of K^+ and A^- is high inside the cell and low on the outside. Because of this concentration difference, a *diffusion force* (F_D) develops which tends to force the ions through the pore (FIG. 5.6a). The A^- cannot leave the cell because of the large size of the protein molecules and therefore it is only the K^+

which move. The result of this egress of positive ions is that the positivity of the cell membrane increases on the outside and decreases on the inside. This produces an increasing electrical force F_E which tends to prevent the further passage of K^+. As the electrical force increases, so the diffusion force decreases and also the resultant force ($F_R = F_D - F_E$). When the two forces become equal, the resultant force becomes zero and the K^+ stop moving (FIG. 5.6b). The electrical potential difference produced is called the resting membrane potential. The resting membrane potential remains constant for as long as the cell is in a resting state. The magnitude of the resting membrane potential varies with different types of cells, but has an average value of 85 millivolt for nerve and muscle cells. The resting state of the cell is also called the polarised state.

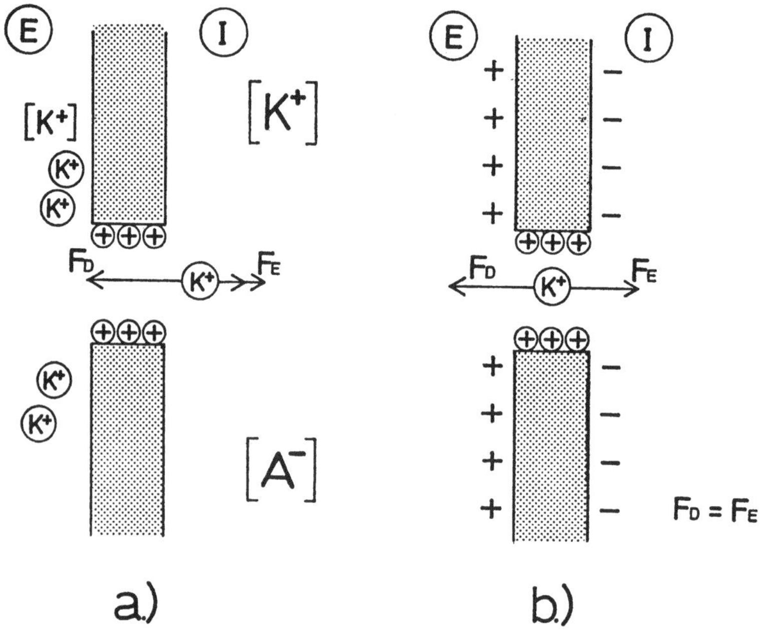

FIG. 5.6 Schematic representation of the origin of the resting potential.

(*a*) A concentration force F_D forces the K+ from the inside of the cell to the outside, and as K+ moves out, an increasing inward electrical force F_E develops.

(*b*) When $F_E = F_D$, the net force acting upon the K+ is zero, and the K+ will therefore cease to move.

Action potential

Under normal circumstances the membrane of the cell has a relatively low permeability for Na^+ ions. When, however, the cell is stimulated by a sufficiently strong electrical current flowing through it, the Ca^{++} is removed from the pores and the permeability for Na^+ increases considerably. The following forces cause movement of the Na^+:

1. diffusion force (F_D) caused by the concentration difference;
2. electrical force (F_E) caused by the resting membrane potential.

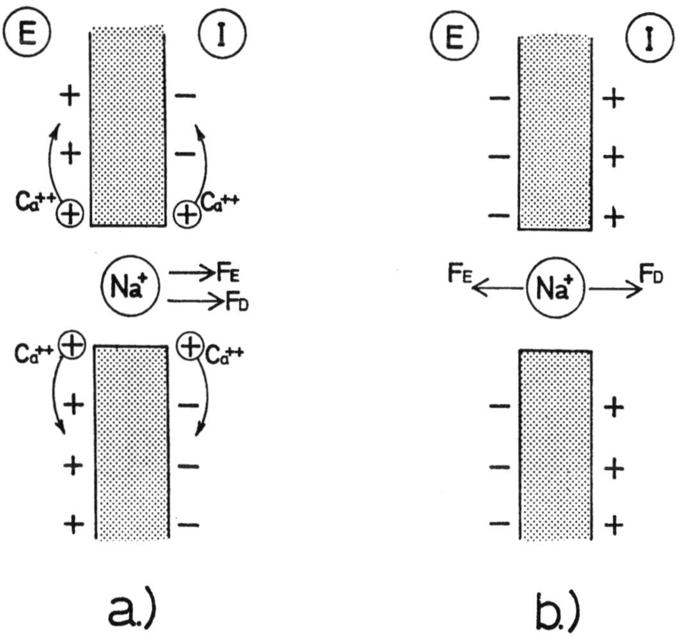

FIG. 5.7 Schematic representation of the depolarization process.

(*a*) When the Ca^{++} are displaced from the pores, Na^+ flows in under the influence of F_E and F_D.

(*b*) The inflowing Na^+ creates a positive charge at the inner surface of the cell membrane; consequently the direction of the electrical force F_E changes, and the inward flow of Na^+ stops when $F_E = F_D$.

The two forces act in the same direction and cause a rapid influx of Na^+. As a result of the influx of positive ions, the negativity on the inside of the cell membrane is not only cancelled, but in fact becomes positively charged as a result of the surplus of Na^+. As the positivity increases so it produces a reversed electrical force that tends to keep the Na^+ out. This means that the two forces acting on the Na^+ oppose each other and when they become equal, the movement of Na^+ will stop. The inside of the membrane has now become positive with respect to the outside, or in other words, the polarity has changed and we say that the cell has become *depolarised.* The process is called depolarisation.

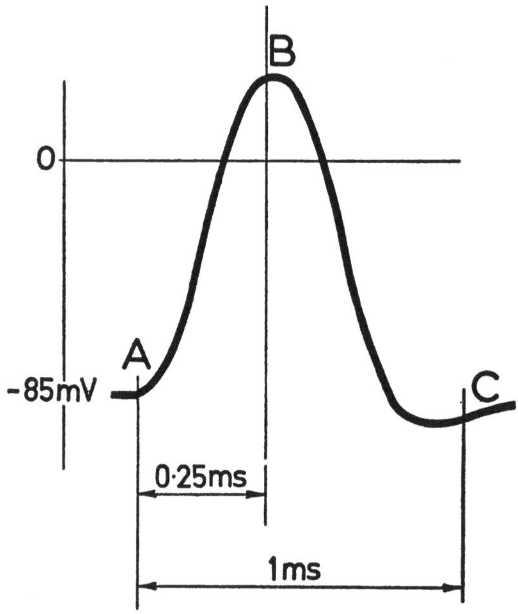

FIG. 5.8 The time course of an action potential in a nerve fibre.
AB = depolarisation. BC = repolarisation.

The new equilibrium reached at the end of depolarisation is, however, unstable and the polarity reverts to the state prevailing before depolarisation, i.e., the resting potential is restored. This process is called *repolarisation.* The time needed for depolarisation and repolarisation of the cell membrane is represented in Fig. 5.8. According to the curve shown in this figure, the cell in question took about 0,25 ms to depolarise (AB) and 1 ms after the beginning of depolarisation it had already regained its resting potential (AC). Repolarisation is mainly the result of an outward movement of K^+ (FIG. 5.9a). The whole potential change (A to C on the curve) is called the action potential (impulse). The excess Na^+ that entered the cell during depolarisation is forced out of the cell by the so-called sodium pump (Chapter 11). For every Na^+ that is pumped out during this process, a K^+ returns to the cell (FIG. 5.9b).

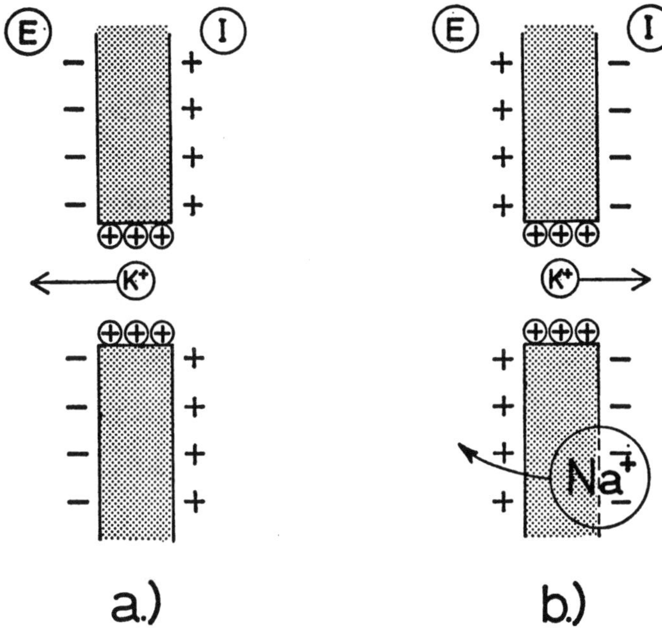

FIG. 5.9 Schematic representation of the repolarization process:
(*a*) outward flow of K+, followed by
(*b*) the removal of Na+ by the sodium pump and a return of some K+

Transmission of the action potential along nerves

Nerve cells are long cells consisting of a cell body and one or more processes. The processes or fibres can attain a length of over a metre, and only a small portion of these can become depolarised at any given moment. The depolarisation of the whole fibre consists of the successive depolarisation of small sections of the fibre (FIG. 5.10). Normally the nerve cell (cell body + processes) is in the polarised state, i.e., the outside of the cell membrane is positive with respect to the inside (FIG. 5.10a). An external stimulus depolarises a small section of the fibre membrane (FIG. 5.10b) and this section becomes positive on the inside and negative on the outside. The rest of the cell remains in the polarised state. Between the polarised and depolarised sections is a dividing line (demarcation line) separating positive and negative areas on the surface of the fibre. The charge difference causes a local current between the two regions. The small adjacent resting region of the cell membrane interprets the local current as an external stimulus and consequently depolarises (FIG. 5.10c). The result is a forward movement of the demarcation line with a positive charge in front of it and a negative charge behind it. The charge pair is called a *dipole*. The transmission of the action potential (depolarisation wave or impulse transmission) can be regarded as the repeated flow of local currents with a subsequent shifting of the dipole.

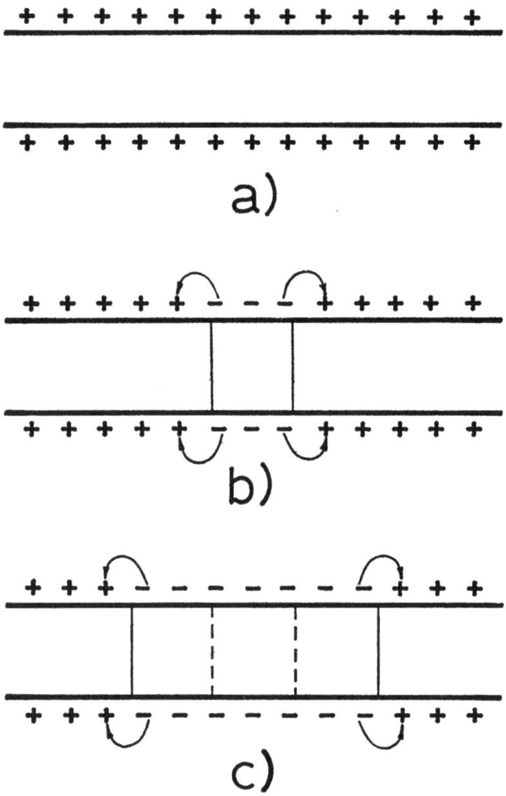

FIG. 5.10 Schematic representation of the conduction of an impulse along a nerve fibre or muscle fibre (see text).

Some nerve fibres are surrounded by a so-called myelin sheath. This special sheath is interrupted at regular intervals to form the nodes of Ranvier (FIG. 5.11). The sheath is an insulator and therefore electrical current can only flow at the nodes of Ranvier. The local current flows between the nodes and therefore the impulse "jumps" from one node to the next. This type of transmission is much faster than that in fibres without a myelin sheath.

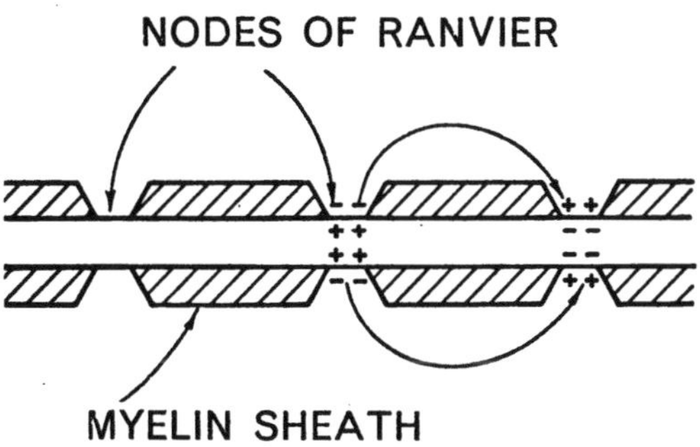

FIG. 5.11 Schematic representation of the conduction of an impulse in a fibre with a myelin sheath.

Resultant dipoles in different tissues

An electrical dipole is composed of a concentration of positive and negative electrical charges closely packed on each side of the demarcation line. The greater the total number of charges alongside each other, the greater the dipole. Nerve and skeletal muscle tissue usually consists of a great number of fibres which do not make anatomic contact with each other. This means that each fibre depolarises and produces a dipole independently. The more fibres that become activated simultaneously, the greater will be the *total dipole*. When nerve fibres reach a muscle, they divide and supply several muscle fibres. The total dipole in a muscle is therefore greater than in a nerve, since the latter contains fewer fibres.

Each muscle fibre receives a branch of a nerve fibre more or less at its centre. This point is called the neuromuscular junction. The nerve impulse is transmitted to the muscle at this point. In the muscle fibre the impulse (action potential, dipole or depolarisation wave) moves in both directions from the neuromuscular junction. The force of contraction of skeletal muscle is determined by the number of fibres that are simultaneously activated. If the amount of work to be performed is small, only a limited number of fibres becomes activated and the resultant dipole is therefore small. If the muscle is performing strenuous work, most of the fibres become depolarised and the resultant dipole is great. In the case of smooth muscle, the resultant dipole is small because the fibres are arranged in an irregular network. Depolarisation of the heart muscle is different from that of skeletal muscle and smooth muscle, because heart muscle is composed of a great number of fibres that *branch and are in direct contact with each other*. The result is that the heart muscle depolarises more or less as a *unit* (because the impulse spreads from one fibre to the next) and the end result is a large resultant dipole. According to the above discussion it is clear that an impulse transmitted in a fibre is accompanied by the movement of a dipole. The dipole is small in nerves, greater in skeletal muscle and greatest in the heart muscle. *Electroencephalography, electromyography and electrocardiography are based on these principles.* In the previous chapter it was shown that when a dipole is produced in a volume conductor, the electricity radiates out to the surface of the conductor. This also holds for the above-mentioned tissues, in which the electricity produced by the moving dipole spreads across the surrounding tissues *to the surface of the body*. By connecting two electrodes on the body surface the electrical changes in the tissues can be measured by a voltmeter.

The electrical activity of the heart

The movement of the dipole in nerves and muscles is relatively simple since it only travels along the length of the fibres. In the heart, the transmission of the depolarisation wave is more complex because the anatomical composition of the heart is so much more complex than in nerves and other muscles.

The pumping action of the heart (FIG. 12). The function of the heart is to pump blood to the lungs and to the tissues of the body. It is made up of two halves functioning together in precisely the same way. Each half consists of two chambers, a receiving chamber, the atrium, and a pumping chamber, the ventricle. The walls of these chambers consist of muscle fibres which contract about once every 0,8 second to perform the pumping action.

The path of the blood through the heart is as follows: blood is received in the right atrium from the whole body, with the exception of the lungs. Blood from the lungs enters the left atrium. From the atria blood flows to the respective ventricles. The atria contract to complete ventricular filling, after which the ventricles contract. The right ventricle pumps blood to the arteries of the lungs and the left ventricle, pumps blood to all other parts of the body via the aorta. The contraction of the chambers thus follows a certain pattern which occurs in a definite time sequence. Consequently the atrial contraction which takes place rhythmically is followed after a determined time by the ventricular contraction. The mechanism which controls

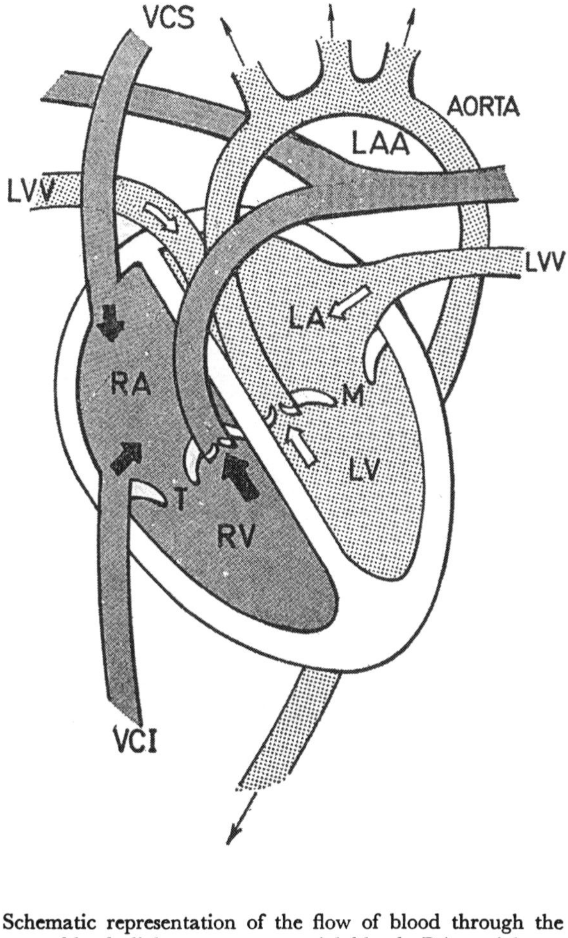

FIG. 5.12 Schematic representation of the flow of blood through the heart. Dark area = venous blood; lighter area = arterial blood; RA = right atrium; RV = right ventricle; LA = left atrium; LV = left ventricle; T = tricuspid valve; M = mitral valve; VCS = vena cava superior; VCI = vena cava inferior; LVV = pulmonary veins; LAA = pulmonary artery.

the rhythmical repetition and sequence of events in the heart, is situated in the heart itself.

Electrical control of the heart (FIG. 5.13). In order to perform work, the heart muscle must depolarise, since depolarisation produces contraction. The heart contains cells known as pacemaker cells, which can depolarise spontaneously. These cells are distributed at specific points in the heart, but the most important are situated in the right

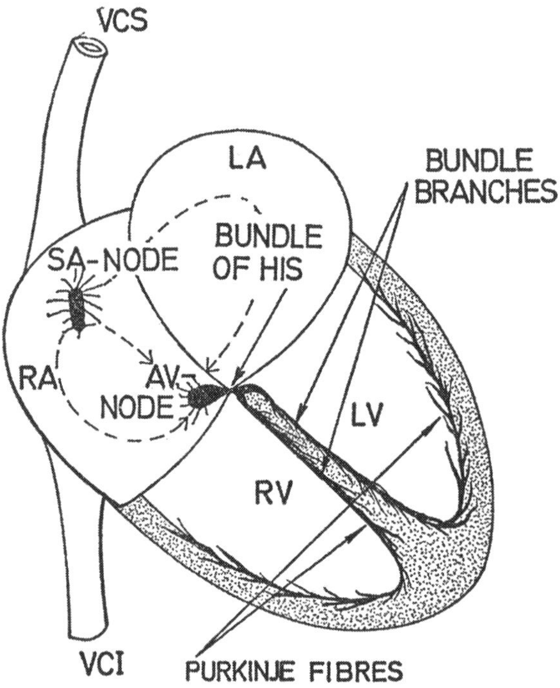

FIG. 5.13 Schematic representation of the tissues responsible for the initiation and conduction of the impulses in the heart (see text). VCS = vena cava superior; VCI = vena cava inferior; RA = right atrium; LA = left atrium; RV = right ventricle; LV = left ventricle.

atrial muscle at the so-called sino-atrial (SA) node. The impulse produced here spreads over the atrial muscle, producing depolarisation which initiates the contraction of the atrial muscle. Simultaneously with the spreading of the impulse in the atrial muscle, it also spreads through special fibres towards a second node, the AV-node, where it is temporarily delayed before it is conducted by a very thin specialised bundle, the bundle of His, to the ventricles. This bundle serves as the only electrical contact between atrium and ventricle. The impulse continues to travel along the special fibres down the wall separating the two ventricles and then spreads over the whole inside of each ventricle. It then enters the ventricular muscle and moves through it from the inside to the outside. When the muscle is depolarised it contracts. The impulse from the SA node therefore depolarises the atrium first, and the ventricles only after the required delay, thus maintaining the correct time sequence of events.

The dipole produced by the heart. During depolarisation of the heart depolarisation of the SA node produces an impulse or depolarisation wave which is transmitted along the atrial muscle, and from there by special fibres to the ventricular muscle. The depolarisation wave can be represented by a moving dipole, the magnitude of which is dependent upon the cross-sectional area of the bundle in which the dipole is moving. The atrial muscle is a thin but quite extensive muscle and the resultant dipole is fairly large. The special fibre bundle has a very small cross-sectional area and the dipole is hundreds of times smaller than in the muscle. The ventricular muscle has the largest cross-sectional area and the dipole is therefore the greatest when it moves in the ventricular muscle.

CHAPTER 6

Variation and Measurement of Electrical Currents

Introduction

The flow of current in a complex electrical circuit is not always as simple as in the case when the circuit is only composed of resistors. Besides resistors, there are other components such as *capacitors*, *coils* and *transistors* that may be included in an electrical circuit and which may influence the current. Combinations of the various circuit components make the construction of a great variety of electronic apparatus possible.

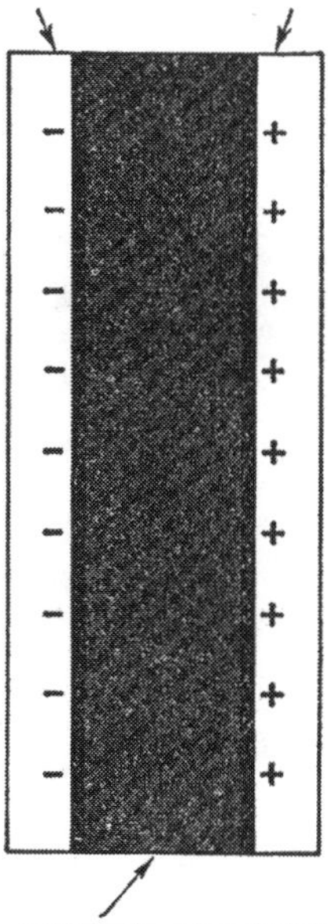

Fig. 6.1 Construction of a capacitor. Two metal plates (electrodes) are separated by a thin layer of insulating material.

Capacitors

The simplest component of a circuit apart from the resistor, is the capacitor. The most important property of the capacitor is its *ability to store energy.* A capacitor is generally composed of two metal plates (electrodes) separated by a thin layer of insulating material (FIG. 6.1).

A charge difference (potential difference or tension) can be created between the two plates. This potential difference can be preserved for a long time, because the opposite charges on both sides of the thin insulating material exert a force of attraction on each other. The process by which the charge difference is created is called *charging*, and it can be carried out with the aid of a charging circuit (FIG. 6.2).

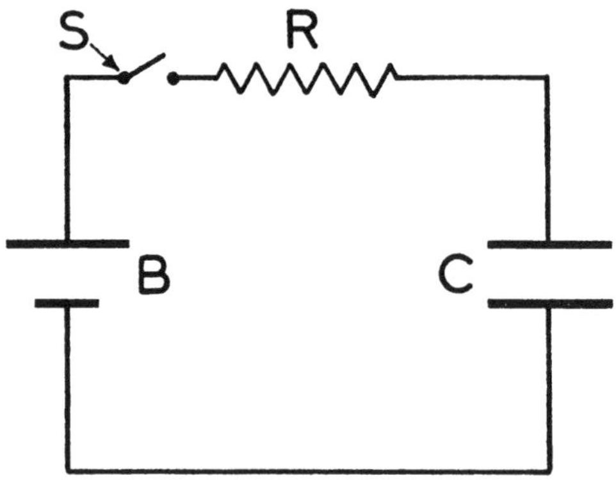

FIG. 6.2 The charge of a capacitor by means of a circuit. B = battery, S = switch, R = resistance, C = capacitor.

A battery (B), a switch (S) and a resistor (R) are connected in series with the capacitor (C) that is to be charged. The battery produces an electromotive force (E) that is kept at a constant level by the chemical reactions in the battery (Chapter 5). Before the charging process starts, the switch is open and the plates of the capacitor are neutral, i.e. the potential difference (V_C) between the plates is zero. The upper plate of the battery has a deficit of electrons in relation to the bottom plate of the battery and therefore also to the bottom and top plates of the capacitor. When the switch is closed, electrons flow from the top plate of the capacitor to the top plate of the battery, in order to cancel out the difference, resulting in the top plate of the capacitor becoming more and more positive (FIG. 6.3). At the top plate of the battery the electrons produce no change in charge, because they are removed by the chemical processes in the battery, in order to ensure that the EMF produced by the battery remains unchanged. This means that the potential difference between the top plates of the battery and the capacitor gradually decreases. The electrical current (charging current) caused by the above-mentioned potential difference decreases as well, and it will stop when the tension on the capacitor (V_C) becomes equal to the EMF of the battery (E), i.e. $V_C = E$.

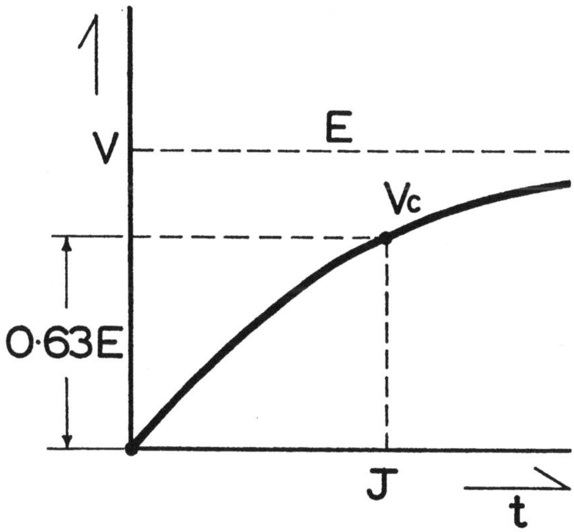

FIG. 6.3 Graphic representation of the relationship between time and capacitor charging. V = voltage; t = time, τ = time constant, E = EMF of the battery, V_C = potential across the capacitor.

The amount of charge that can be stored in a capacitor is dependent upon the *physical properties* of the capacitor on the one hand and upon the *tension* (V_C) to which the capacitor becomes charged, on the other. For a given capacitor, the charge, therefore, is only dependent on the tension and is proportional to it, i.e.,

$$Q = CV_c$$

Q = charge in coulombs
V_C = tension on the capacitor in volts
C = constant for a given capacitor, called the *capacitance* of the capacitor. Unit is the *farad* (F).

The capacitance of a capacitor is determined by the physical properties of the capacitor and can be calculated from the following formula:

$$C = \frac{KA}{d} \text{ in farads}$$

A = surface area of the plates in metres
d = distance between the plates in metres
K = constant, depending on the type of insulation material used. It is also called the *dielectric constant.*

The equation $Q = CV_C$ can also be written as follows:

$$\frac{Q}{V_c} = C$$

This means that the charge per volt in a given capacitor is constant, and is equal to the capacitance of the capacitor.

Time constant

As the capacitor is charged the potential difference between the battery and the capacitor gradually decreases (FIG. 6.3). This means that the charging current and thus the charging rate also decreases. It therefore takes very long before the capacitor becomes fully charged. The initial charging rate, however, is fast and about two thirds of the tension is reached after a relatively short time. It has been found that the time necessary for a certain resistor-capacitor combination to reach 63% of its full tension is constant. This constant is called the *time constant* (τ) of the resistor-capacitor combination. Numerically, it is equal to the product of capacitance and resistance, i.e., $\tau = RC$

τ = time constant in seconds
R = resistance in ohms
C = capacitance in farads

Problem 6.1. What is the time constant for the combination of R = 1 kilo-ohm and C = 1 microfarad?

$$\begin{aligned}\tau &= RC\\ &= 10^3 \times 10^{-6}\\ &= 10^{-3}\text{s}\\ &= 1\text{ ms}\end{aligned}$$

Problem 6.2. Assume that the EMF of the battery connected to the combination in Problem 6.1 is 20 V, what is the tension of the capacitor after 1 ms?
1 ms is the time constant, τ (from Problem 6.1). After the time τ, 63% of the full tension is reached, i.e. $0{,}63 \times 20 = 12{,}6$ volts.
When the charged capacitor (FIG. 6.2) is disconnected from the battery, it retains its charge. This charge can now be used for other purposes by discharging the capacitor. *Discharging* means the equalising of electrical charges on the two plates, and it can be achieved by

connecting a resistor between the plates. During discharge the potential difference between the plates decreases gradually (Fig. 6.5) and the current does too, since it is proportional to the potential difference. It takes, therefore, a very long time before the capacitor is completely discharged. As in the case of the charging of the capacitor, the time taken for it to discharge is also constant. The *time constant* is again determined by the product of the capacitance and the resistance of the discharging circuit ($\tau = RC$).

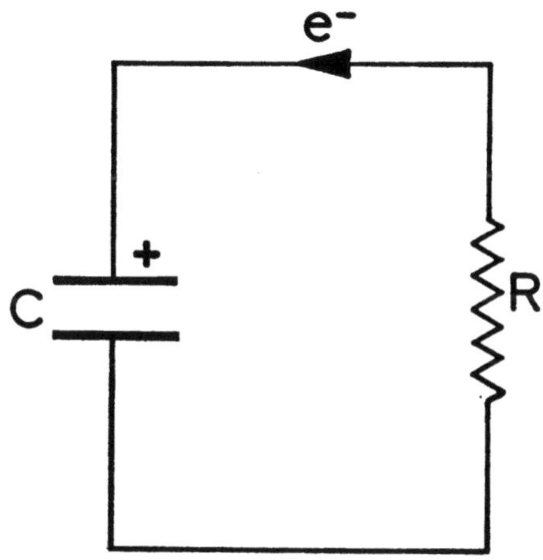

Fig. 6.4 Schematic representation of the circuit for the discharge of a capacitor C by a resistance R. Electrons (e—) flow from the negative to the positive electrode of the capacitor.

According to this equation, it is clear that the progress of a dynamic process, i.e., the flow of charging current, is determined by the physical properties of the capacitor and the resistor. For this reason a resistor-capacitor combination can be used as an *electronic timer*. To understand the principles of an electronic timer, one can compare it with a sand timer. In the sand timer the sand takes a certain time to flow from the upper part of the timer to the lower part. The time taken is constant and is dependent upon the amount of sand that must flow and the ease with which it flows through the connecting portion. The same applies to the electronic timer, except that here one is concerned with the flow of electrons from one electrode to the other. The time taken for a capacitor to discharge to a certain level (37%) is dependent on the capacitance of the capacitor and the ease with which electrons move in the wire (resistance). In the case of the sand timer, the time taken for the upper portion to run dry is noted by a person keeping constant watch on the device. In the

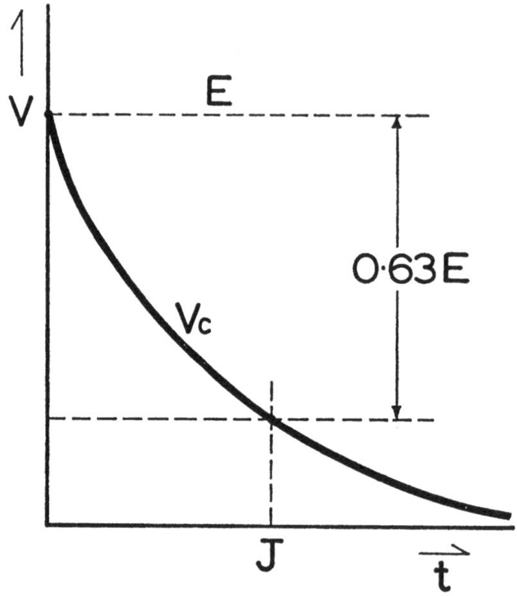

FIG. 6.5 Graphic representation of the relationship between time and capacitor discharging. V = voltage, τ = time constant, t = time, E = voltage before discharge begins, V_C = potential across the capacitor.

case of electronic timers, however, special instruments observe the potential changes which are an indication of the time that has elapsed during the change in level of potential. These instruments can be used to activate switches for the switching on or off of apparatus or circuits.

An example of the use of a capacitor-resistor combination is in the defibrillator. This apparatus is used to terminate fibrillation of the heart (a serious irregularity of the heart rhythm). It causes the flow of a certain amount of electrical charge through the heart of the patient. The time of flow of charge is important, because if it is too short, the defibrillator is inefficient, whereas if it is too long it can cause permanent damage to the heart. In practice a charged capacitor is connected to the patient. Since the patient is a volume conductor, his body presents a certain resistance, and the time taken for the capacitor to discharge is determined by the capacitance of the capacitor and the resistance of the patient.

Electromagnetism

The force that exists between two magnetic poles has already been mentioned in Chapter 3. It has been shown that like poles repel one another and unlike poles attract each other. Magnets are usually made out of a special iron alloy, but can also be made from normal iron by winding a coil around the iron and passing an electrical current through the wire of the coil (FIG. 6.6). Such a magnet is called an electromagnet. The polarity of an electromagnet is determined by the direction of the current flowing in the wire. For example, if the magnet is so orientated that one end faces the observer and the other end is away from him, the pole closest to him is the south pole of the magnet if the current is flowing clockwise in the wire. The strength of the electromagnet is determined by the physical properties of the coil and the metal on the one hand, and by the current flowing in the wire on the other. By physical properties one understands the number of turns in the coil, the dimensions of the coil and the type and dimensions of the iron core. The iron core can actually be omitted altogether, but then the electromagnet will be very weak. In a given electromagnet therefore, the strength of the magnet is determined by the *current* flowing in the wire and is proportional to the current, i.e.,

$$M = kI$$

M = magnetic strength in SI units

k = constant for a given electro-magnet

I = current flowing in the wire in amperes

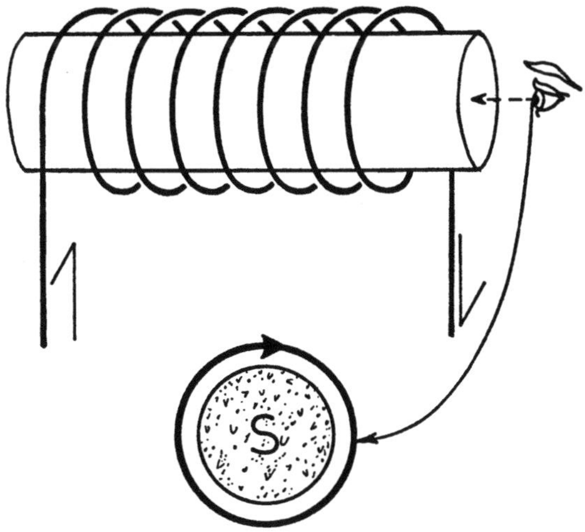

FIG. 6.6 Schematic representation of an electromagnet, the current flows clockwise in the coil and the south pole (S) is nearest to the observer.

An electromagnet behaves in the same way as an ordinary magnet, i.e. a force which develops between an electromagnet and an ordinary magnet may be represented by the following equation (Chapter 3):

$$F_M = K_M \frac{M_O M_E}{d^2}$$

F_M = magnetic strength in newtons
M_O = strength of ordinary magnet
M_E = strength of electromagnet
K_M = constant

The above-mentioned two equations provide the basis upon which electromagnetic measuring instruments have been developed. The magnitude of the force can easily be determined by measurement of the movement caused by the force. The force is directly proportional to the strength of the magnets, the strength of the one being constant and the strength of the other being directly proportional to the current flowing in the coil. The movement caused by the force is therefore proportional to the current flowing in the wire, and can thus be used to indicate the amplitude of current in the wire.
In FIG. 6.7 a simple measuring instrument is illustrated schematically. It consists of a permanent horse shoe magnet and a small electromagnet that can rotate about an axis at its centre. When current flows in the coil a N-S pole pair is produced in the electromagnet. Forces then develop between the two magnets, and these forces are determined by the strength and direction of the current flowing in the wire. The small electromagnet rotates as a result of the forces. A small spring (not shown in the diagram) is stretched during the rotation and produces a force that opposes the rotation. When the opposing force becomes the same in magnitude as the force causing rotation, the pointer will come to rest. The extent of angular displacement of the pointer therefore indicates the current flowing in the wire. When the current is flowing from A to B the upper pole is the N pole and the rotation is clockwise. This can be called the positive direction of the current. When current is flowing in the opposite direction, the upper pole becomes the S pole and the pointer moves anti-clockwise, indicating a negative current. The strength of the current is read off the scale. Notice that the instrument measures current, and therefore it is called an *ammeter*. We know, however, from Ohm's law that there exists a simple relationship between current and potential difference, and the ammeter can thus also be used as a *voltmeter*, provided that the resistance in the coil is known.

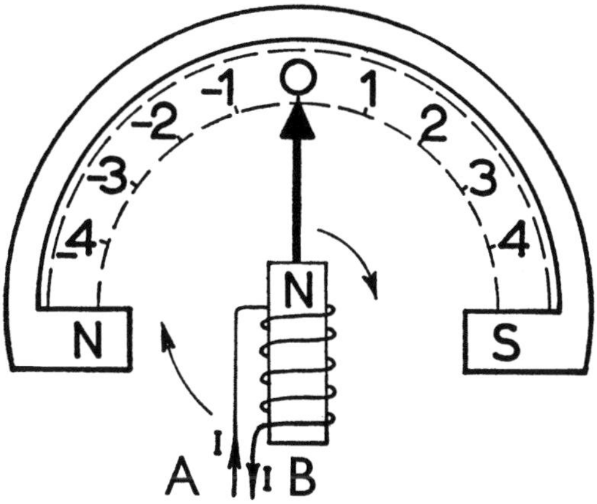

FIG. 6.7 Schematic representation of a simple electrical instrument (see text).

Graphic recorders like those used in the electrocardiograph are also voltmeters. In these cases the pointer of the meter is replaced by a pen that writes on a moving paper strip.

Production of an alternating current

In the previous discussion it was shown that the interaction between a magnet and the current flowing in a coil causes angular displacement of the coil, i.e., *electricity* + *magnetism* = *angular displacement.* The extent and direction of angular displacement is determined by the strength and direction of the current. If a coil rotates between the poles of a magnet, an electrical potential difference is produced in the coil: *magnetism* + *angular displacement* = *electricity.* The magnitude and the polarity of potential difference is dependent on the angular displacement of the coil.

Since the displacement of the coil changes constantly with rotation, the potential difference alternates as well, i.e. it increases and decreases in two directions, first in the one, then in the other. Because of this alternating potential difference, the current produced is called an *alternating current* (AC). The stronger the magnet, the greater the number of turns in the coil; and the faster that the coil rotates, the greater is the maximum potential difference. Alternating current for everyday use is provided by huge *generators* at power stations, because it is easier to produce alternating current than direct current on a large scale. The power needed to turn the big coils is provided by steam engines. Alternating current has another advantage in that it can be increased or decreased by means of *transformers.*

Power stations produce potential differences of hundreds of kilovolts and the wires that conduct this electricity are called high tension lines. The electricity is conducted to the cities where it is transformed to tensions of 220 or 250 volts before being used.

Semiconductors

Semiconductors are materials which, by virtue of their conducting properties fall between conductors and insulators. There are two types of semiconductor materials. One type has an excess of electrons and is called the negative or N-type, the other has a deficit of electrons and is called the positive or P-type semiconductor.
By combining the two types, the electrical current can be varied in different ways. For example a N-type and a P-type semiconductor can be combined to construct a diode. A *diode* is a circuit component that allows current to flow in one direction only. When the positive pole of a battery is connected with the positive pole of the diode, the current will flow. If it is connected the other way around, however, there will be no flow of current. Diodes are also used as *electronic switches* and *rectifiers*.
Another example of semiconductor combinations is the *transistor*, in which semiconductors are joined together in combinations of PNP or NPN. Transistors are used as *amplifiers* of electrical current or potential difference. In the case of the heart the electrical potential difference which must be measured is in the order of 1 mV. This is too small a value to be registered. In order to make registration possible, the potential difference must be amplified 1 000 times.

The oscilloscope

An oscilloscope is a special voltmeter based on principles, different from those underlying the electromagnetic voltmeter. It is used to produce a *temporary graph* on a screen. In Fig. 6.8 A is a *metal*

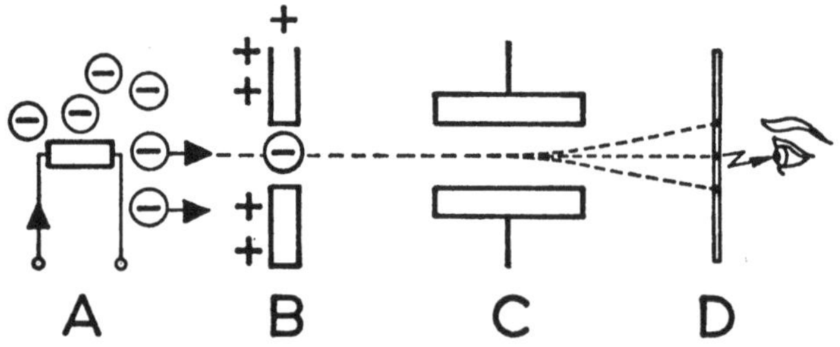

Fig. 6.8 Schematic representation of the most important components of an oscilloscope. A = metal element heated by electricity which then releases electrons; B = positive plate (electrode) which accelerates the electrons released by A; C = deflection electrodes; D = screen.

filament which, when heated, emits electrons. B is a positive plate (*accelerating plate*) which attracts the electrons, thereby causing their acceleration. There is an opening in the accelerating plate which allows the electrons to pass through. There are two so-called *deflection plates*, C, and a phosphorescent screen which emits light rays when the electron beam impinges on it.

The electron beam moves between the deflection plates. If there is no potential difference between the plates, they will not influence the movement of the electrons and therefore the electron beam falls on the centre of the screen. When the two deflection electrodes are charged positively and negatively respectively, they exert an influence on the moving electrons. The positive electrode attracts them, while at the same time the negative electrode repels them. Since the electrons are moving at great speed, they are only slightly deflected from their linear path. If the top plate is positive, the electrons impinge on the screen above the centre, and if the bottom plate is positive, they reach it below the centre. The deflection on the screen is dependent upon the potential difference between the plates and can thus be used to give information about an unknown potential difference connected to the plates. The apparatus can for example be set to show a deflection of one centimetre for a one millivolt potential difference. When this change is effected, the light point on the screen moves vertically.

To produce a graph the light point must move horizontally along the

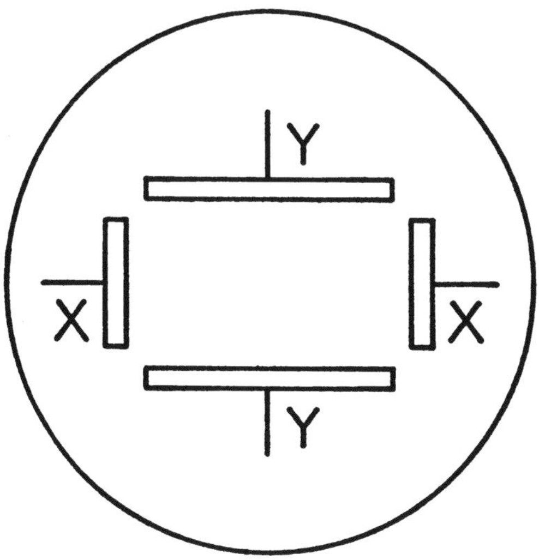

FIG. 6.9 The deflection electrodes as seen from the front: XX = electrodes for horizontal deflection; YY = electrodes for vertical deflection.

screen at a constant rate. The horizontal movement can be obtained by including two more deflection plates perpendicular to the original ones. The potential difference between these last two plates is positive on the righthand side and increases gradually. The light

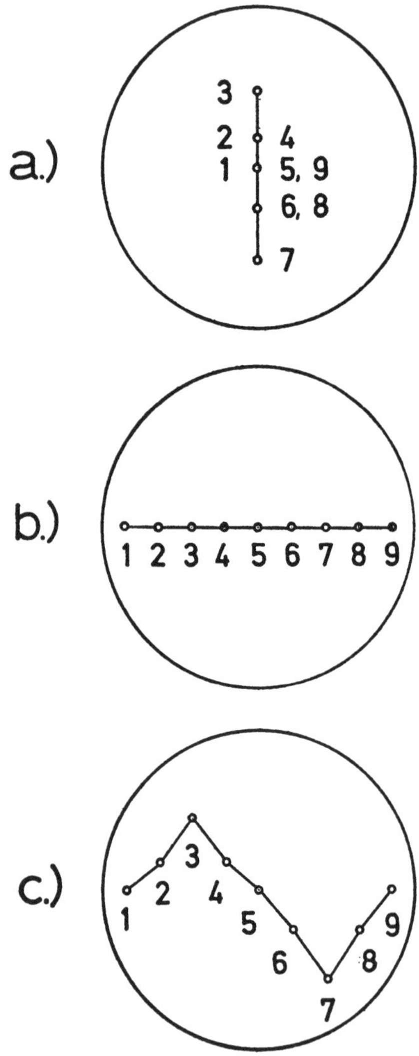

FIG. 6.10 The functional mechanism of the deflection electrodes in an oscilloscope: (*a*) only the vertical electrodes functioning; (*b*) only the horizontal electrodes functioning; (*c*) both electrodes functioning simultaneously.

point moves slowly from left to right on the screen. When the light point reaches the righthand side, the left plate suddenly becomes positive and the light point moves quickly over to the left, and the process repeats itself. The rate at which the light point moves is dictated by the rate at which the potential changes. The apparatus can for example be set in such a way that the light point moves 10 cm per second on the screen. In this case each centimetre is equivalent to 0,1 second (FIG. 6.9).

The graph produced on the screen of the oscilloscope is the total effect of the two movements. The *horizontal deflection* is a constant movement and indicates time. The *vertical deflection* is dependent on the potential difference which it indicates. FIG. 6.10 depicts the result of the vertical and horizontal deflections. The oscilloscope is often used in medicine. It has an advantage over graphic recorders in that it can continuously draw graphs which remain on the screen for a short time only. It can thus be used for the continuous monitoring of the ECG of a patient. When an oscilloscope is used to observe the electrocardiogram, it is called a cardioscope.

CHAPTER 7

Bio-Electronic Equipment

Introduction

A great variety of apparatus is used in medicine and biology. *Apparatus which functions in direct contact with living organisms is called bio-electronic apparatus.* Apparatus such as theatre lights, airconditioners and sterilisers also play an important role in medicine and in biological research but, because they do not come into direct contact with the living organisms, they are not considered to be bio-electronic apparatus.
Bio-electronic apparatus can be classified according to the *application*, the *type of body function* with which it is concerned and the *principles of operation.*

1. On the basis of its application, bio-electronic apparatus can be divided into (*a*) *diagnostic apparatus*, (*b*) *therapeutic apparatus*, (*c*) *surgical apparatus* and (*d*) *research apparatus.* Diagnostic apparatus is used to establish whether a biological system is functioning normally or abnormally. This apparatus plays an important role in the examination of patients. The electrocardiograph and the blood pressure meter are examples of this type. Therapeutic apparatus is used to correct a malfunction in the body. A pacemaker and a respirator belong to this group. Surgical apparatus corrects or helps to correct an anatomical deviation. The heart-lung machine and surgical diathermy equipment are of this nature. Research apparatus is used to find out new facts about biological systems.

2. Bio-electronic apparatus can be divided into (*a*) *primary* and (*b*) *secondary* apparatus, depending on the nature of the body function being measured. Primary apparatus is concerned with the inherent electricity of the body. Examples of this type of apparatus are the electrocardiograph and the pacemaker. Secondary apparatus is used to monitor or regulate the non-electrical functions of the body, the blood pressure meter and the respirator being examples.

3. On the basis of functional principles, bio-electronic apparatus can be divided into (*a*) *input*, (*b*) *output* and (*c*) *compound apparatus.*

It must be remembered that *each bio-electronic apparatus can be classified according to all three methods of classification.* For example an electrocardiograph is a primary diagnostic input device, and a respirator a secondary therapeutic output device.

Classification of bio-electronic apparatus on the basis of functional principles

As far as the flow of current is concerned, every bio-electronic apparatus has an input and an output (FIG. 7.1).

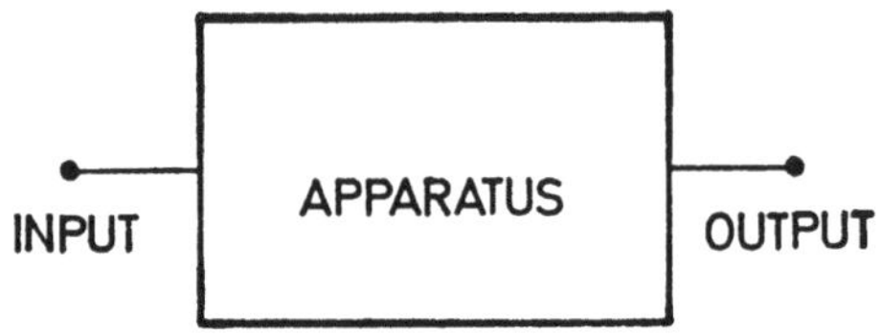

FIG. 7.1 Schematic representation of a bio-electronic apparatus.

Electricity produced outside the apparatus reaches the apparatus at its input. On the other hand, the electricity produced in the apparatus itself leaves it at the output. When an apparatus is connected to a biological system via its input, it is called an *input device* (FIG. 7.2a). When, however, it is connected to the biological system via its output, its called an *output device* (FIG. 7.2b). It is possible to make a connection at both ends which is the case with compound apparatus. According to the primary source of electricity, compound apparatus can further be divided into two groups. If the biological system is the source of electricity, and electricity reaches the apparatus at its input and leaves it at the output and subsequently passes back into the biological system, the apparatus is called an *input-output* apparatus (FIG. 7.2c). On the other hand if the electricity is produced in the apparatus, and leaves it by the output and after passing through the body it re-enters the apparatus at its input, it is called an *output-input* apparatus (FIG. 7.2d).
The advantages of the classification into input and output groups can be readily seen. Each group functions according to certain principles common to the apparatus in the group and the different apparatus in the group vary in fine detail only. The compound apparatus comprises one or more units of each of the single groups, and its working principles and problems are those of the simple groups combined. An additional poblem arises, however, concerning the connection between the parts forming the compound apparatus.

Input devices

In the case of input devices the *electricity is produced by the biological system* and enters the apparatus at its input. Input devices are mostly used for diagnostic purposes, e.g. the electrocardiograph, electro-

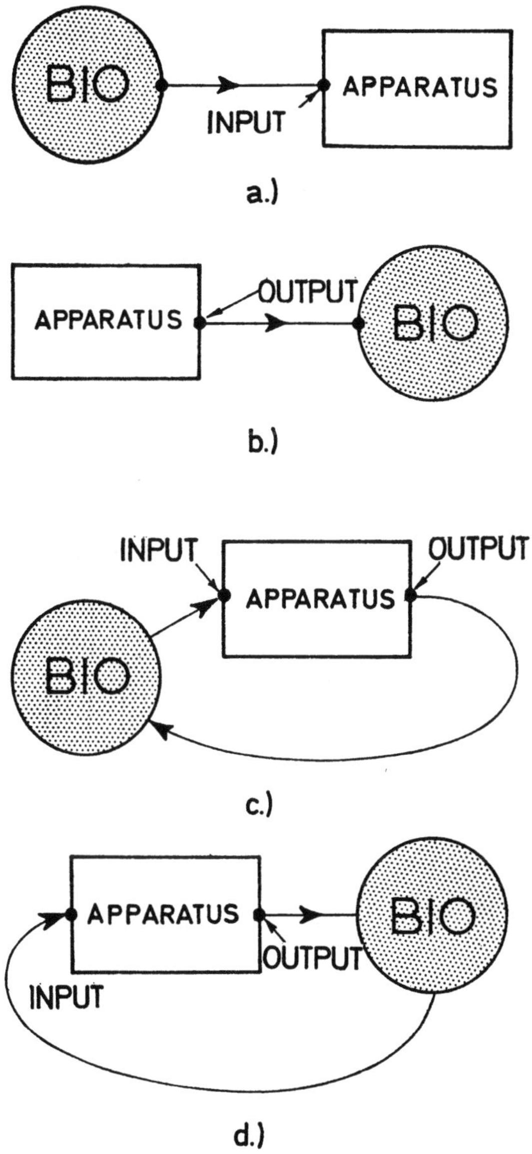

FIG. 7.2 Classification of bio-electronic apparatuses based on functional principles (see text).

myograph and electroencephalograph. Some input devices, however, are used for therapeutic purposes, for instance, artificial limbs and motivation apparatus.

In general the principles upon which input devices work can be described as follows (Fig. 7.3): a part of the biological system, such

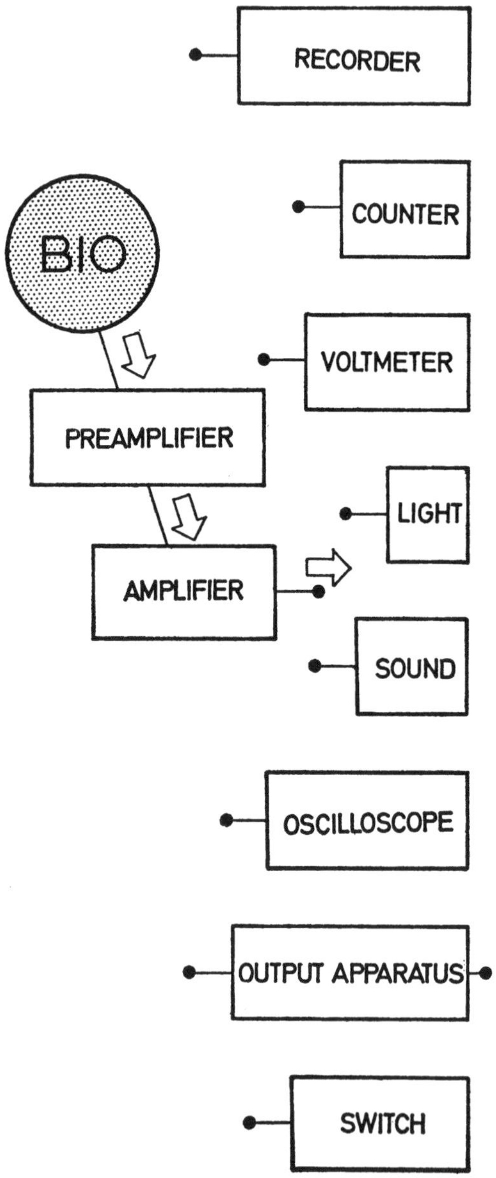

Fig. 7.3 A block diagram of an input apparatus.

as a nerve or muscle, produces an electrical impulse, which in fact behaves like a moving dipole (Chapter 5). The electrical field produced by the moving dipole extends even to the surface of the body. The changing potential differences (signals) can be picked up by the apparatus. The connection between the biological system and the artificial apparatus is achieved by *electrodes*. One end of the electrodes is connected to the biological system, the other to the electrical apparatus. Secondary input devices differ only from the primary apparatus in that the electrodes are replaced by *transducers*, that convert the physical signals to electrical signals.

Before a signal can be received by the electrical apparatus the electrical connection between the electrode (or transducer) and the biological system must be correct. This rather difficult concept can perhaps be better understood with the aid of a simple analogue. Let us imagine that water must flow from one tank to another, and that the output opening of the first is large in diameter and the input opening of the second is small in diameter. In order to ensure that no loss of water occurs while it is being transferred, a converter must be included which adapts the large diameter to the small diameter. In the case of bio-electronic apparatus, electricity flows in the place of water, from a biological system to an electrical system. The important parameters that have to be adapted between the two systems are not physical characteristics such as diameter, but electrical characteristics such as resistance and capacitance. This adaptation is performed by the *pre-amplifier.*

The signal provided by the electrodes is usually too small to be detected and must therefore be amplified. A part of this amplification is usually performed by the pre-amplifier and the rest is done by the *main amplifier*. The amplified signal can be used in different ways. It can be recorded *permanently on paper* or *temporarily by an oscilloscope.* The signals *can be counted* over a time period, to give an indication of the frequency of the activity. The impulses or the absence of impulses can operate a *loudspeaker*, a *flashing light* or a *switch* or they can be fed into an output apparatus thus forming a compound apparatus.

All input apparatuses are built up from the above-mentioned basic components. In the various types of apparatus the characteristics of the components are different, because the electricity produced by the different organs of the body, or the electricity derived from the various transducers, is different. There are three characteristics that must be considered in this regard, namely the *type of electrode or transducer used*, the *amplitude* of the *electrical signal* available at the input of the apparatus and the *form of the electrical signal.* The first two characteristics determine the *type of pre-amplifier* that has to be used and the third determines the required *frequency response* of the apparatus.

We have seen that different tissues produce dipoles which differ from each other in amplitude. Heart muscle, for example, produces 1-2 mV, skeletal muscle 0,05-0,2 mV (that can under certain extreme activity increase to as much as 2 mV) and nerve fibres 0,005-0,1 mV, on the surface of the skin (FIG. 7.4). Let us assume that a recorder needs 1 volt for 1 cm deflection of the recording pen. If we want the 1 mV signal of the heart to cause 1 cm deflection, it must be amplified by a thousand times. The 0,2 mV of skeletal muscle will produce 1 cm deflection if amplified five thousand times, while the signal produced by the nerves must be amplified ten

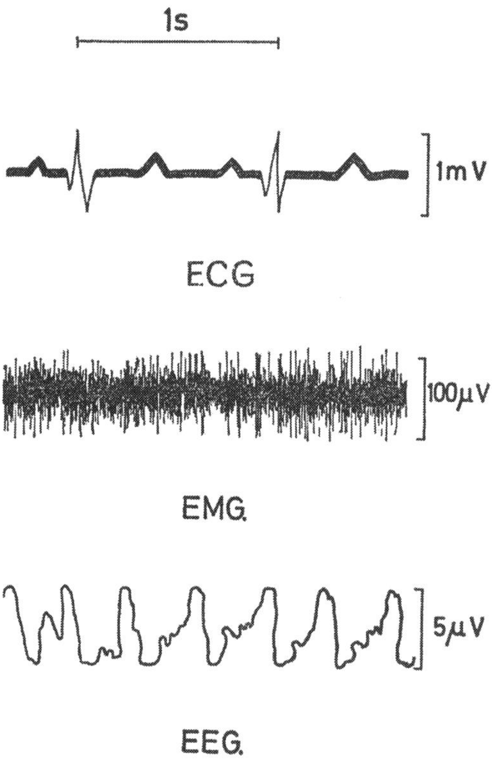

FIG. 7.4 Registration of electrical changes in the heart (the electrocardiogram, ECG), in skeletal muscle (the electromyogram, EMG), and in the brain (electroencephalogram, EEG).

thousand fold to achieve the same effect. The amplification of the pre-amplifier in the various types of apparatus, must therefore be different. The electrodes and transducers that are used for the various purposes differ in their electrical characteristics. The electrical characteristics of the pre-amplifier must therefore be suited to the purposes concerned.

When an electrical change takes place in a biological system, it must immediately be recorded by the apparatus, and the graph that is registered must be a faithful time record of what is happening. The faster the rate of change of the electrical signal, the more difficult it will be to ensure that all sections of the apparatus are faithfully following it. For each apparatus there is a maximum rate of change at which the recording will still be reliable. It must therefore be ensured that the apparatus can follow the electrical signal which it is supposed to measure. The electromyograph that records rapidly-repeated impulses must have a much larger range of response than the electrocardiograph that records much slower impulses.

Generally speaking one might say that all apparatus should be built to give an accurate reproduction of anything that may occur. In practice this is not done, because unwanted signals can be eliminated by limiting the response of the apparatus. This can be illustrated with the following example: when the activity of the heart is registered in a person at rest, the recording shows five consecutive deflections repeated with each heart beat (FIG. 7.5a). The first of the deflections is the P-wave, the next three form the QRS-complex and the last is the T-wave. The P-wave and the QRS-complex represent the movement of the dipole in the atria and the ventricles respectively, and the T-wave represents the recovery of the ventricles.

If the patient moves and the apparatus has an extended response capacity, the electrical impulses produced by the skeletal muscle will also appear on the registration (FIG. 7.5b). By limiting the response to the slower heart impulses the registration can be made "clean". Apart from the unwanted muscle signals, various other electrical disturbances can be produced or picked up by the system. This so-called "noise" can also be eliminated by the restricted response of the apparatus.

Most input devices are used for diagnostic purposes, for instance the electrocardiograph is used as an aid in the diagnosis of certain heart diseases and the electromyograph as an aid in diagnosing certain nerve and muscle abnormalities.

Artificial limbs and motivation apparatus are examples of therapeutic input devices. Patients who have lost their lower arm can be fitted with an *electro-mechanical arm* that enables them to perform certain limited movements. This operates on the principles which follow. The upper arm still has muscles under nervous control. The electrical potential produced by these muscles is picked up and operated upon in such a way that it can be used to control the artificial arm. In this way the mechanical arm is under direct control of the brain.

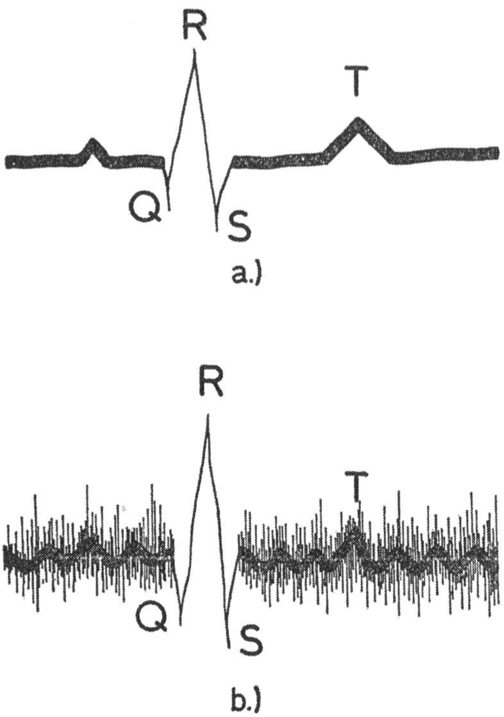

FIG. 7.5 (*a*) A normal registration of the electrical changes in the heart.
(*b*) The same registration as in (*a*) but including interference.

Motivation apparatus is used for the promotion of nerve regeneration. Damaged nerves regenerate if their activity is preserved. In the early stages of regeneration muscular activity is absent. This activity is what the patient hopes to see. A sensitive electrical apparatus can, however, register the early reappearance of activity before the patient can see it. This electricity produced can be used to switch on an apparatus such as a sewing machine or a lathe in which the patient is interested. The patient must produce a minimum nerve activity in order to work the above-mentioned apparatus. As the nerves regenerate, their activity increases, and the detectable electricity is increased. The patient is thereby motivated to send out more and more impulses and thus indirectly assists with the regeneration of his damaged nerves.

Output devices

In output devices the electricity is produced in the device and lead to the patient. Most output devices are used for therapeutic purposes, i.e. they are used to correct or to compensate for a malfunction in the body. There are however, a few output devices which are used for diagnostic purposes, such as stimulators used by physiotherapists to investigate nerve function.

Output apparatus consists of a *generator* which produces the electrical impulses, and an *output converter* which ensures that the form and the amplitude of the impulses are optimal, and that the apparatus is suited to the biological system (FIG. 7.6). Examples of apparatus belonging to this group include pacemakers, defibrillators and apparatus used for electronarcosis and electroshock. The fundamental principles which underline the operation of all output devices are the same for this whole group. The *frequency*, *form* and *amplitude* of the electrical impulses which are delivered to the biological system, are determined by the defect that has to be corrected. The pacemaker and the defibrillator may be mentioned as examples of this type of apparatus.

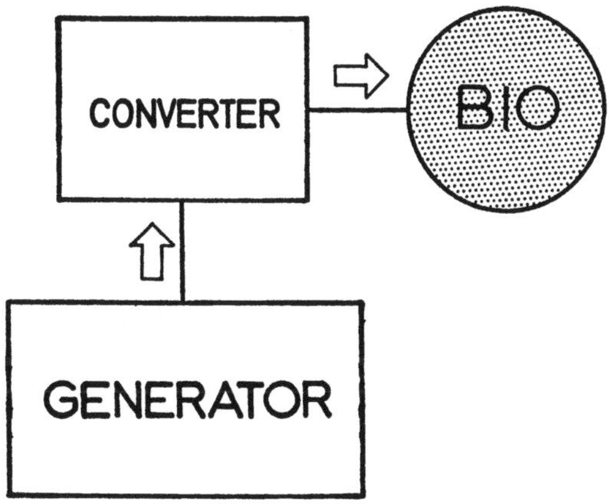

FIG. 7.6 A block diagram of an output apparatus.

Electrical pacemakers and defibrillators

An electrical *pacemaker* is used to maintain the activity of the heart in patients suffering from *complete heart block*. It has already been mentioned (Chapter 5) that during the normal activity of the heart the impulses are produced in the SA node (which is thus the normal pacemaker of the heart) and these are transmitted through specialised tissue, the atrio-ventricular bundle, to the ventricles. The thin atrio-ventricular bundle (Bundle of His) is the only functional connection between the atrial and ventricular muscle entities. In complete heart block the conduction along the specialised atrio-ventricular bundle is interrupted, and consequently the atria and ventricles contract independently of one another. The abnormality can be demonstrated on the electrocardiogram (FIG. 7.7). Impulses are still produced in the SA node, at a frequency of about 70 per minute, and on the ECG normal P waves can thus be seen. The atrial muscle is thus functioning normally. The impulses do not reach the ventricles, however, these being controlled by other parts of the special conducting tissue in the ventricles, which act as pacemakers. These so-called ectopic pacemaker cells produce impulses at a much slower rate, completely independently of the atrial rate. Because the ventricular contraction rate is slow, the volume of blood pumped by the heart may be decreased to such an extent that the tissues become deficient in oxygen. The brain is especially sensitive with regard to oxygen deficiency, and consequently the person may faint.

To alleviate this condition regular electrical impulses must be conducted through the heart muscle at the normal rate. This can be achieved by an electrical pacemaker which produces electrical impulses at a rate of 70 per minute. Each impulse should have a duration of 2-6 milliseconds and an amplitude of 2-5 volts. The generator produces the impulses which are then changed to the right form and amplitude by the converter, which also ensures that the electrical characteristics of the apparatus and the biological system are suited to each other.

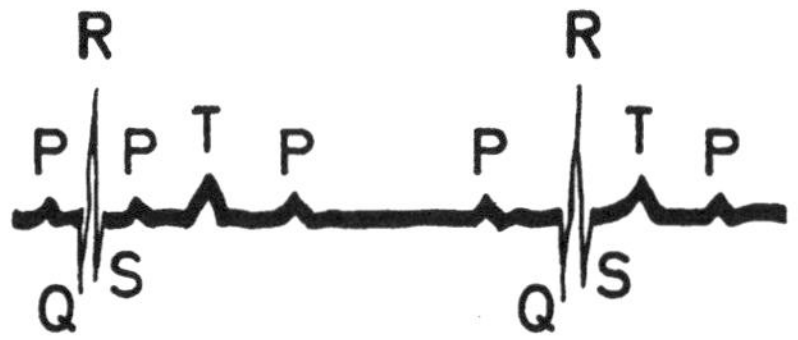

FIG. 7.7 Electrocardiogram of a patient with total heart block. Notice the dissociation of the P-waves and the QRS-complex.

It must be mentioned that despite the fact that the ventricles now contract at the right frequency the problem is not completely solved since the atria and the ventricles receive their impulses from independent sources. Normally the ventricular rate is controlled by

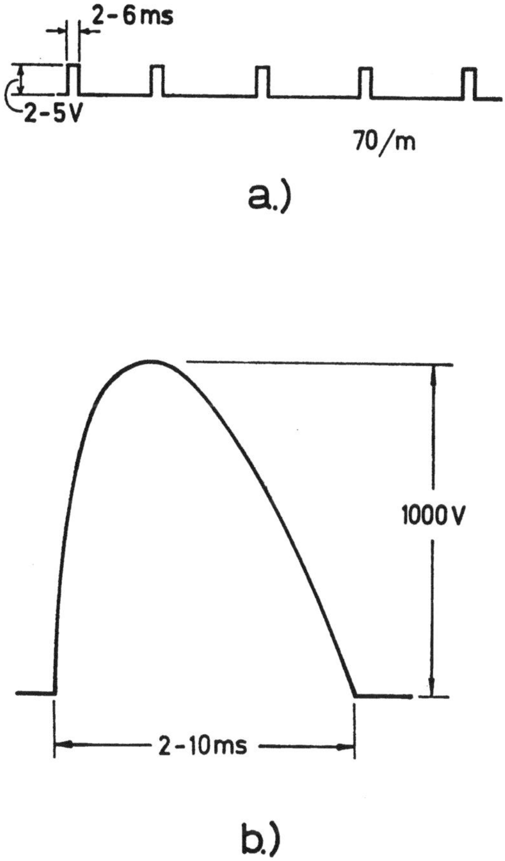

Fig. 7.8 Electric shocks supplied by (*a*) a pacemaker, and (*b*) a defibrillator.

impulses originating at the SA node in such a way that atrial contraction precedes ventricular contraction. This permits the atrial contraction to complete ventricular filling before the ventricles contract. Although atrial contraction is not essential for the filling of the ventricles, it is responsible for about one third of this. If atrial contraction does not precede ventricular contraction by the delay required to permit filling, each ventricle becomes a weaker pump by 30% and the blood flow to the tissues is reduced. Another problem that exists with an artificial pacemaker is the inability to increase the heart rate when the person is more active. For example the heart rate of a resting person is 70 beats per minute but it can increase to 140 per minute during strenuous exercise. This is a very important functional adaptation of the heart which enables it to pump sufficient blood to the tissue, thereby supplying the necessary oxygen and nutrients and removing the waste products of metabolism. Artificial pacemakers on the other hand provide impulses at a constant rate irrespective of the degree of physical activity.

The problems of (a) insufficient synchronisation between atrial and ventricular contraction and (*b*) inability to adapt the activity of the heart to the activity of the body have been largely eliminated by synchronised pacemakers.

A *defibrillator* is used on patients whose hearts *fibrillate.* In fibrillation the various muscle fibres in the atrial or ventricular muscle contract asynchronously. According to which of these is concerned we speak of atrial or ventricular fibrillation. In the latter instance the cardiac function is seriously disturbed and consequently the patient dies unless fibrillation is terminated at once. This can be done by means of a defibrillator. In the case of atrial fibrillation the patient can lead a useful life for years without defibrillation.

The first half of the graph in FIG. 7.9 shows the electrocardiogram of a patient with ventricular fibrillation. At moment S, defibrillation is brought about by the passage of a strong electrical shock current through the whole heart. The intensity of the shock varies between 1 000 and 2 000 volts and it has a duration of 2-10 milliseconds (FIG. 7.8b). As it has been mentioned before (Chapter 6), this electrical shock is obtained by discharging a capacitor.

The differences between pacemakers and defibrillators can be summarised as follows.

1. A pacemaker stimulates the heart rhythmically through electrodes connected to the heart muscle. Only a single muscle fibre has to be stimulated and the action potential is transmitted to the

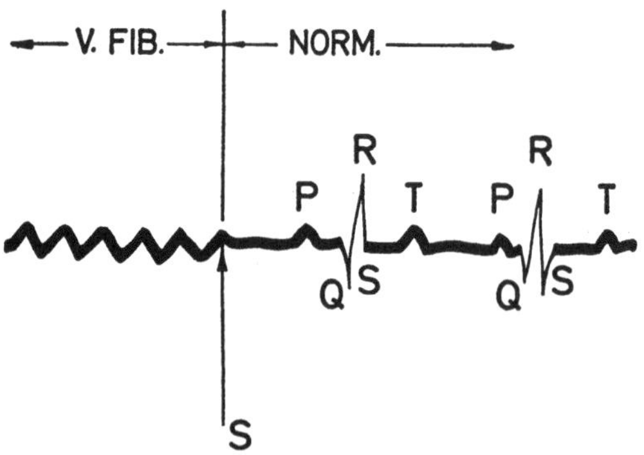

FIG. 7.9 Electrocardiogram of fibrillating ventricles and in which the fibrillation was terminated by a shock applied at S.

whole heart muscle by the muscle fibres themselves. The impulses applied are small and are repeated at the normal frequency of the heart.

2. The defibrillator stimulates the whole heart simultaneously and once only. This is done by a severe shock through two large electrodes which are brought in contact with the heart or with the chest wall.

Compound apparatus

Compound apparatus can be divided into two groups, viz, input-output apparatus and output-input apparatus.

1. *Input-output apparatus.* In this case electricity is produced in the biological system and fed into the apparatus at its input (FIG. 7.10). The impulses are changed by the apparatus and are conducted through the output back to the biological system. Synchronised pacemakers and defibrillators are examples of this type of equipment.

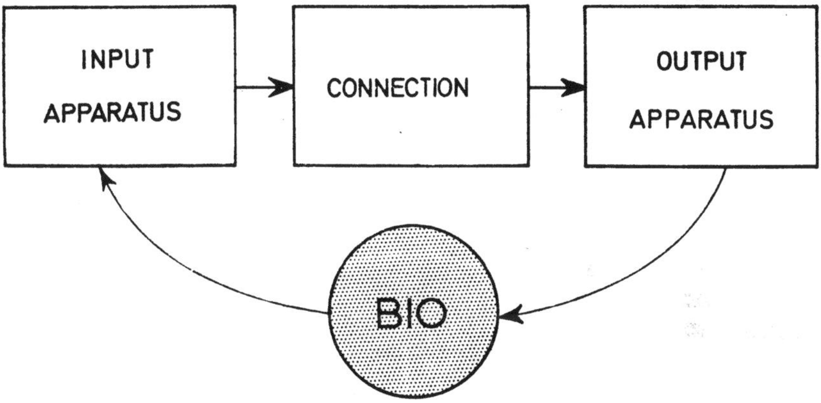

FIG. 7.10 A block diagram of an input-output apparatus.

2. *Output-input apparatus.* Diagnostic stimulators belong to this group. The electricity is produced in the apparatus, taken to the biological system and picked up again by the apparatus.

The shortcomings of normal pacemakers can be overcome by *synchronised pacemakers*. With these the ventricles are stimulated at the right time interval after atrial contraction. Special electrodes pick up the impulses which are produced in the atria and transmit them to the input apparatus. This is actually an ECG apparatus, but instead of passing into a recorder the impulse passes to a special time delay circuit. From here the impulses are used to operate a pacemaker. The impulses are then conducted from its output to the heart. In this way the ventricles are paced according to impulses from the atria and thus the normal filling of the ventricles is ensured.

Synchronised Defibrillators are used in patients with atrial fibrillation. During atrial fibrillation some irregular atrial impulses reach the ventricles. As a result of fibrillation the atria do not function as a pump and their contribution to ventricular filling is thus lost. Persons suffering from this complaint are able to live a quiet life, but are usually unable to be very active. The ECG shows more or less normal QRS-complexes and T-waves, but at an irregular rate (FIG. 7.11).

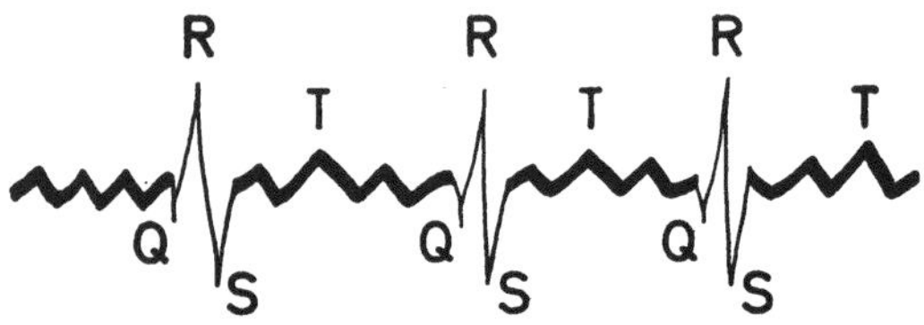

FIG. 7.11 Electrocardiogram of a heart with fibrillating atria (atrial fibrillation).

Just as in the case of ventricular fibrillation, this condition can be terminated by an electric shock. It is, however, very important that the shock is not applied during the T-wave. If this happens,

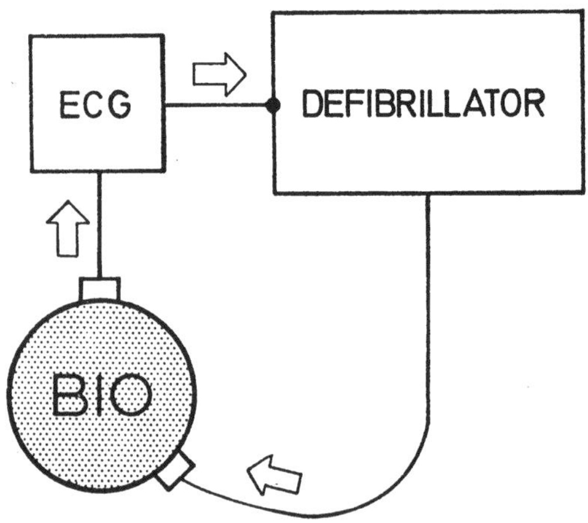

FIG. 7.12 A block diagram of the synchronized defibrillator.

fibrillation of the whole heart, including the atria and ventricles, can be precipitated. For this reason the period of time during the T-wave is called the vulnerable period. To make sure that the shock does not occur in the vulnerable period, a synchronised defibrillator is used. The QRS-complexes of the electrocardiogram are used to activate the defibrillator. Applying the shock immediately after the QRS-complex ensures that it does not occur during the vulnerable period. FIG. 7.12 represents the block diagram of the apparatus and FIG. 7.13 shows electrocardiograms before and after a shock is applied.

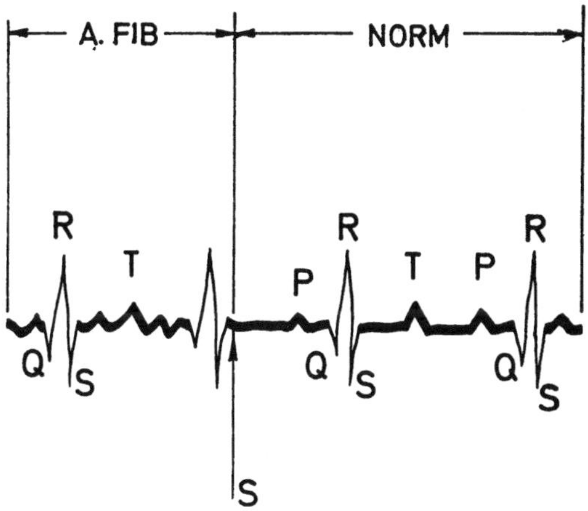

FIG. 7.13 Registration of the electrical changes which take place in the heart during atrial fibrillation. Fibrillation is terminated by applying a shock at S.

SECTION III

LIQUIDS, GASES AND HEAT

CHAPTER 8

Common Properties of Liquids, Gases and Heat

Introduction

Liquids, gases and heat are discussed in the same section since there is a physical relationship between them as well as a similarity in their dynamic behaviour.*

Physical relationship. Matter can exist in three states: solid, liquid and gaseous. The state of matter at any particular moment in time is determined by its temperature.

Dynamic behaviour. The dynamic behaviour of liquids, gases and heat is similar to that of electricity. The flow of electricity and the factors which influence it are given by Ohm's law (Chapter 4), according to which current is proportional to the potential difference between the two ends of a conductor, i.e.

$$I = \frac{V}{R}$$

I = current intensity
V = potential difference
R = resistance to the flow of current

The resistance is determined by the physical properties of the conductor as follows:

$$R = \frac{\rho l}{A}$$

where

ρ = resistivity of the conductor material
l = length of the conductor
A = cross-sectional area of the conductor

Ohm's law is also applicable to the flow of liquids, gases and heat, but potential difference is replaced by pressure difference in the case of liquids and gases, and by temperature difference in the case of heat. The formula for the determination of the resistance to the flow of heat is the same as that for electricity, but is slightly different for

*N.B. — Liquids and gases are matter and heat is molecular movement.

liquids and gases, where the cross sectional area term "A", is replaced by "A^2". The numerical value of the constant is obviously different in each case.

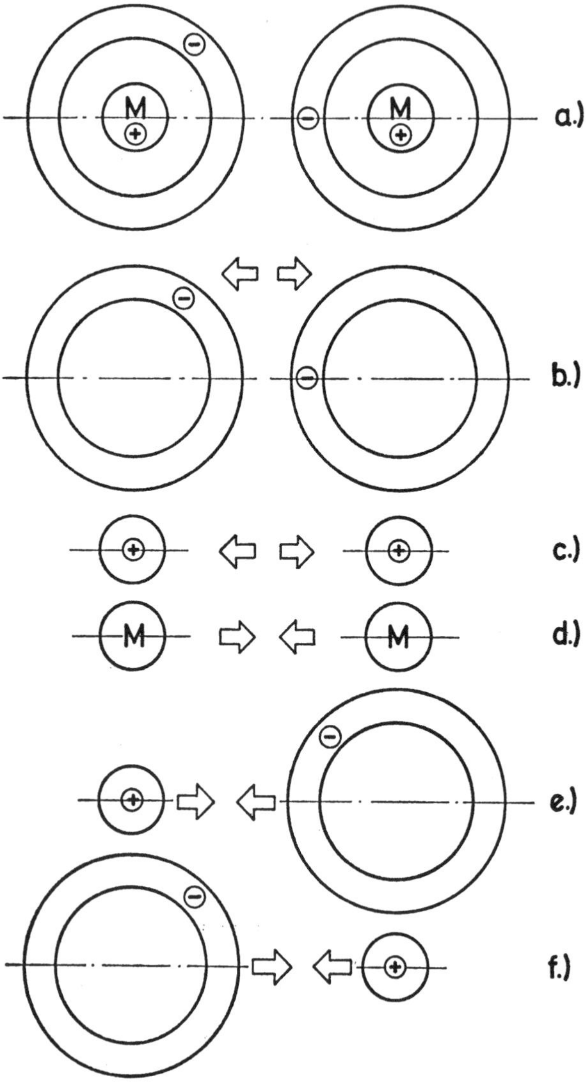

FIG. 8.1 Intermolecular forces. (*a*) Atoms consist of nuclei and electron orbits. There are forces of repulsion, viz, (*b*) between two orbits, (*c*) between two nuclei; and forces of attraction, viz, (*d*) between two masses, and (*e*) and (*f*) between a nucleus and electron orbit.

Liquids and gases exert a pressure caused by gravity. These pressures and ways of measuring them, are discussed in this chapter.

States of matter

Matter can exist in three states depending on its temperature. Water, for example is a *solid* below 0°C, a *liquid* between 0°C and 100°C and a *gas* above 100°C. In order to understand the various states of matter, we must study the so-called "*Kinetic Theory*" which explains these states in terms of molecular movement. Molecules of materials in all three states are in continuous movement, brought about by the continuous change in the equilibrium between the forces of attraction and repulsion acting on the molecules. Each atom consists of a positive part, the nucleus, and a negative part, the orbiting electrons (FIG. 8.1*a*). There exist therefore various forces of attraction and repulsion between the molecules. There is, for example gravitational force between two nuclei (FIG. 8.1), and an electrical force of attraction between the nuclei of one molecule and the electrons of another (FIG. 8.1 *e* and *f*). There are also forces of repulsion between two nuclei and between two electron orbits (FIG. 8.1*b* and *d*). The electrons are in continuous movement in their orbits (Chapter 4), and continually changing their distribution in the orbits. This means that the electrical forces and consequently the equilibrium between the molecules is continuously changing. The final result is a continuous oscillation of molecules.

In solids, the nuclei are close to each other, which means that the attraction forces (cohesive forces*) are great. Solids actually have a fixed lattice structure and molecular movement is small. On heating, the movement of the molecules in the lattices become greater and greater and the cohesive forces weaker and weaker. The molecules move away from each other and the material expands. When the temperature reaches the melting point, the movement is so violent that the lattice structure becomes destroyed and the state changes from solid to liquid. The change of state requires heat energy and the temperature stays constant during the change (FIG. 1.3). When all the molecules of the material have changed their state from solid to liquid, further heat energy will again increase the molecular activity (temperature) and cause further expansion. In the liquid state the cohesive forces are sufficient to keep the molecules within the liquid mass, but are insufficient to ensure a definite form. At the boiling point of a material the influence of the cohesive forces diminishes to such an extent that the molecules leave the liquid

*Forces of attraction between molecules of the same material are called cohesive forces, and those between molecules of different materials adhesive forces.

mass and appear in the gaseous form. In this case energy is also necessary for the change of state. In gases, repulsion forces are primarily present, and these tend to disperse the molecules as much as possible. This is the reason why the molecules of a gas tend to fill the container completely.

The role of heat in determining the state of matter should be clear from the foregoing discussion. Heat increases the molecular activity within the temperature limit of a certain state, and at the temperature limit it changes the state of the material. At the melting and boiling points the heat is used to destroy certain forces of cohesion without any rise in temperature. For this reason the heat supplied is called *latent heat* (latent = hidden). The change in volume that accompanies the change in temperature beyond the melting and boiling points can be used for the measurement of temperature as in the case of the thermometer (Chapter 2).

In summary we may say that matter exists in three states, viz, solids that have a definite form and volume, liquids with a definite volume but no definite form and gases having neither a definite form nor a definite volume. Liquids and gases are also called *fluids* because they are capable of changing their form.

The flow of fluids and of heat

Flow is movement, and it requires a force to produce it (Chapter 3). In the case of electricity there is electromotive force caused by charge or potential difference. In the case of fluids and heat, forces are caused by pressure and temperature differences respectively. Ohm's law, which describes the relationship between factors determining the flow of electricity, can also be applied to fluids, if the potential difference (V) is replaced by pressure difference (P), viz,

$$I = \frac{P}{R}$$

I = flow intensity

P = pressure difference

R = resistance

In the same way Ohm's Law can be used to describe the flow of heat, if the potential difference is replaced by temperature difference (T), viz,

$$I = \frac{T}{R}$$

I = flow intensity

P = temperature difference

R = resistance

For heat the form of the equation for the determination of the resistance is precisely the same as for electricity:

$$R = \frac{kl}{A}$$

k = *specific resistance* to the flow of heat

l = length of the conductor

A = cross-sectional area of the conductor

In the case of fluids flowing through a tube the equation becomes:

$$R = \frac{\eta l}{A^2}$$

η = viscosity of the fluid

l = length of the tube

A = cross-sectional area of the tube in which the fluid is flowing

N.B.—The difference between the two expressions is that in the case of fluids the *square* of the cross-sectional area is used in place of the cross-sectional area.

In order to understand the connection between resistance and the flow of fluids in tubes it is necessary to understand the nature of flow. There are two types of flow: laminar and turbulent. Slow flow as it is found mostly in the human body is laminar. Turbulent flow is faster and is found only in a few places in the body. In the case of laminar flow, the fluid moves in concentric cylindrical layers that slide over each other. In a certain layer all the molecules of particles move with the same speed, but the speed decreases in the layers further and further from the centre. In the case of turbulent flow, the flow is irregular and cross currents are produced causing the molecules to collide with each other and with the walls of the tube. If the cross currents become too violent, the flow will produce a murmur.

The above-mentioned expression for resistance is only valid in the case of laminar flow. Friction that develops between adjacent layers as they slide over each other is called *viscosity*, which is a quality of the fluid itself. In the case of a thick fluid, the layers slide over each other with difficulty, and the viscosity is high. In thin fluids the opposite pertains. Viscosity is usually denoted by η (Greek letter eta = η) and the unit in which it is measured is the *poise*. The viscosity of a fluid is often compared with that of water at 20°C in which case one talks of relative viscosity. The relative viscosity ot blood at body temperature is for example 3,5-5,0.
Forces of attraction between the molecules of a solid and a fluid are called adhesive forces. When a fluid is flowing in a tube, the forces of adhesion between the tube and the outside fluid layer are so great that this layer remains stationary. This means that the flow is actually retarded by the adhesive force. In the case of electricity the rate of movement of the electrons is the same over the whole cross-sectional area of the conductor, because there are no adhesive forces. This is the reason why the cross sectional area of the tube through which the fluid flows has a greater influence on the magnitude of the resistance than is the case in electricity, and that is the reason why the cross sectional area is squared. This means that the cross-sectional area of the tube plays a greater role in the determination of the resistance than any of the other factors. The equation shows that doubling of the cross-sectional area reduces the resistance to flow four-fold.
By way of summary it can be said that the following factors influence the flow of electricity, heat and fluids:

1. potential, temperature and pressure differences;
2. the dimensions of the conductor, viz, length, cross-sectional area;
3. the specific resistance or viscosity.

In the case of fluids a fourth factor, namely the square term, is added.

Application of the laws of flow to the human body

The knowledge of the above-mentioned principles regulating the flow of fluids and heat is important in understanding the flow of blood in the circulatory system, the flow of air in the respiratory system and the regulation of body temperature.
Circulatory system. In the circulatory system the blood is forced through the vessels by pressure differences, caused mainly by elastic forces in the vessels. As a result of the pressure differences in the various sections of the circulatory system, the blood flows through the tubes (the vessels) to the different organs. The quantity of blood flowing through an organ in a unit of time is dependent on the pressure

difference between the two ends of the tubes and the resistance of the tubes. The resistance is determined by the cross-sectional area of the vessels, the length of the vessels and the viscosity of blood. The cross-sectional area can be changed by the contraction or relaxation of muscles in the walls of the vessels. Through this mechanism the distribution of blood to various organs can be varied.

As has been mentioned, the resistance decreases for example, four times if the cross-sectional area of the blood vessel doubles, and therefore with the same pressure applied the flow increases four times. A small change in cross-sectional area causes a big change in the blood flow to an organ. There are many factors that can influence the muscles in the walls of the vessels and thereby their cross-sectional area, for instance, nerve impulses, hormones such as adrenalin, noradrenalin and thyroxin, other chemicals, like carbon dioxide, oxygen and histamine, and physical factors like temperature. With the aid of these factors the blood flow to the various organs can be regulated according to the requirements of the organ at any given time. In cases where the blood vessels are constricted as a result of anatomical changes, due for instance to calcification of the vessels (atherosclerosis), the blood flow to the affected organ becomes disturbed.

As has been mentioned, viscosity is a property of the fluid, which in this case is blood. At body temperature, the relative viscosity of blood is about 3,5-5,0 and this is mainly determined by the quantity of red blood cells and plasma proteins in the blood. In persons who are anaemic, i.e., who have too few red blood cells, the viscosity is low, and their blood flows easily. On the other hand, if the blood contains too many red blood cells, as in a condition called polycythaemia, the viscosity is increased and the resistance to flow becomes great. Further application of the laws of flow to the circulatory system are discussed in Chapter 10.

Respiratory system. The second system in which the laws of flow play a role is the lungs. During inspiration the pressure in the lungs decreases below atmospheric pressure and air flows in. During expiration the pressure in the lungs increases above atmospheric pressure and air flows out. The flow of air in the air passages is controlled by pressure differences and the resistance of the air passages. If the air passages become constricted, as in asthma, or in tumour of the air passages, the resistance increases to a great extent, causing serious respiratory disturbances. To improve the flow of air in these cases the pressure differences can be increased by artificial respirators. The respiratory system and artificial respirators are discussed in greater detail in Chapter 9.

Temperature regulating system. Life is the result of a series of complicated chemical reactions, that can only proceed at a certain constant temperature. The reactions themselves produce heat energy that flows through the body and is also yielded up by the body to its surroundings. In both instances the flow is regulated by the laws previously discussed. The temperature regulating mechanisms of the body are discussed in Chapter 12 in detail.

Pressures in fluids caused by gravitational forces

Pressure can be defined as the amount of force exerted per unit area. If an object weighing 10 N stands on the ground, it produces a pressure on the ground. If the weight of the object remains unchanged, but the area on which it stands on the ground becomes smaller, the force per unit area increases. Let us assume that the bottom area of the above-mentioned object is 5 m^2, the pressure that it exerts on the ground is then $^{10}/_5 = 2\ P_A$; if the area is only 1 m^2, the object exerts a pressure of $^{10}/_1 = 10\ P_A$. Pressure is therefore an indication of the *intensity* of the force that is exerted on the structure of the ground.

The importance of pressure can be demonstrated by using a simple example. If a five cent coin is pressed against a drawing board with a great force, it only leaves a mark corresponding to its edges. If, however, a drawing pin is pressed against the board it goes deeply into it. The force applied has not changed, but it has been concentrated in a much smaller area, i.e., the pressure has increased. Other examples that illustrate this phenomenon are for instance when one tries to cut bread with the blunt and the sharp edges of a knife, or the difference between marks left by wide and stiletto heels on floors.

Pressure plays an important role in materials where the molecules move in relation to each other. In fluids the movement of molecules takes place as a result of the force exerted by the molecules on each other. Pressure in a fluid gives an indication of the intensity of the force acting on individual molecules.

When a person is in a lying position, the weight of the body is not distributed over the whole surface, but is concentrated on certain anatomical points. At these points the blood supply, and therefore nourishment, is reduced. This is the reason why patients who are immobile and who for long periods lie in the same position, develop *pressure sores*. These can be prevented by regular changing of position of the patient, or by the use of special matresses. These matresses are made in such a way that various parts can be alternately inflated and deflated, thus changing the pressure points. Other applications of the principles of pressures in medicine is in the use of hypodermic needles and knife blades. The sharper the point of a needle and the cutting edge of a blade, the less pressure is needed to penetrate the tissue, and consequently the extent of disturbance of tissue is reduced. Gravitational force is the most common force causing pressure; it draws materials towards the centre of the earth. The pressure caused at a certain point in a fluid is determined by the weight of the fluid

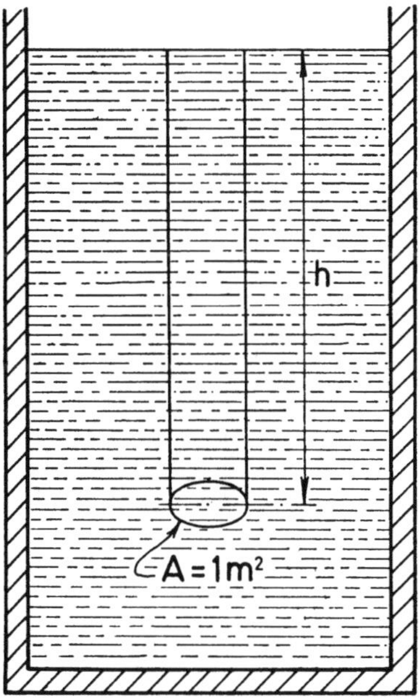

FIG. 8.2 Pressure in a fluid is caused by the weight of the fluid column above a unit area (A); h is the height of the column.

above a unit area at the point (FIG. 8.2). The mass (M) of the fluid above a unit area around the point is the product of the volume (V) of the fluid column above the area and the density (ρ) of the fluid (Chapter 3), i.e., $M = V \times \rho$. The volume of the column is given by the product of the area (A) of the bottom surface and the height (h) of the column, i.e., $V = A \times h$. The formula for mass is then $M = \rho Ah$.

The weight of the column reflects the force of attraction of the earth and is given by $F_G = gM$, where g is the gravitational constant (Chapter 3). If mass is replaced by ρAh the expression for weight becomes:

$$F_G = g\rho Ah$$

Pressure (P) is force (weight) divided by area, i.e.,

$$P = \frac{g\rho Ah}{A} = g\rho h$$

With the aid of this expression the pressure at a point in a fluid can be determined if the density of the fluid and the depth of the point in the fluid are known. The pressure in fluids is called *hydrostatic pressure.*

Every object on earth is subjected to air pressure caused by the weight of the air above it. The air pressure is greater at sea level because the height of the air column is greater. For the same reasons the air pressure is less on mountains as there the column of air is smaller. In the case of gases, however, additional pressures exist which are caused by the movement of the molecules. These pressures will be discussed in the following chapter.

An important property of liquids is the fact that their upper surface is always horizontal. This means that all points on the surface are at the same distance from the centre of the earth. This property has a practical application. Let us assume that we have two containers in which the level of water is not the same (FIG. 8.3). If the two containers are joined by a tube, water will flow from container A, which has the higher level, to container B, which has the lower level. Flow will continue until the two levels in the containers are the same.

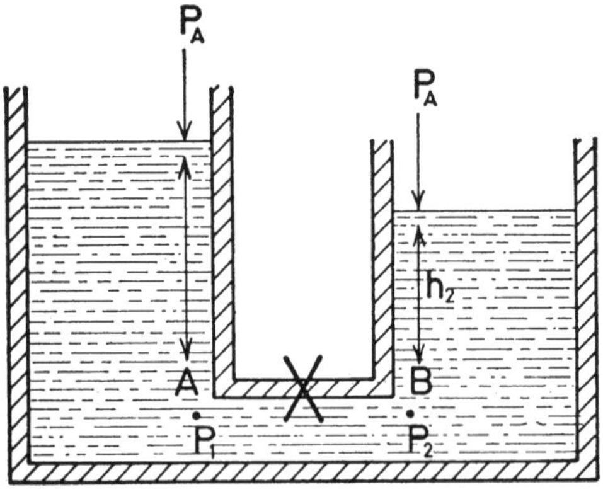

FIG. 8.3 The flow of a liquid takes place from point A to point B, because the Pressure P_1 at point A is greater than the pressure P_2 at point B. PA is the atmospheric pressure.

The pressure is produced by the weight of the columns above the two ends of the tube joining the two containers. Apart from this the air above both containers produces a pressure. The total pressures above the two ends are P_1 and P_2 respectively, and are given by

$$P_1 = P_A + \rho g h_1 \text{ and } P_2 = P_A + \rho g h_2$$

P_A = the air pressure above the container
h_1 and h_2 = the height of the two columns, respectively
ρ = density of the liquid
g = gravitational constant

According to the above equations the pressure difference between the two ends of the connecting pipe is

$$P_1 - P_2 = P_A + \rho g h_1 - (P_A + \rho g h_2)$$
$$= \rho g (h_1 - h_2)$$

As a result of this pressure difference, flow will take place and continue until the pressures at the two ends of the tube, i.e. in the two containers, become equal, or in other words when the two columns

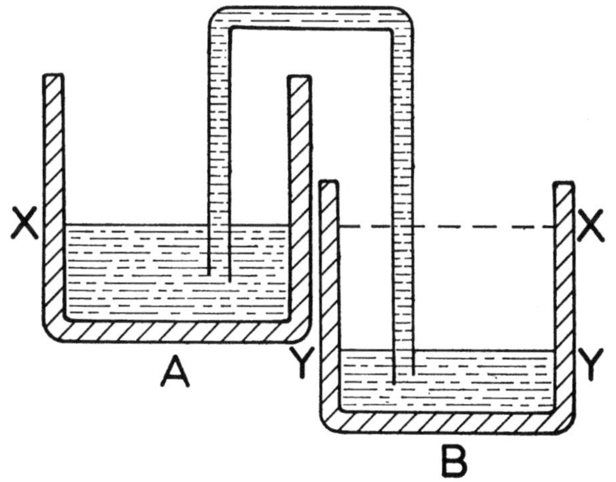

FIG. 8.4 The flow of water from one container A to another B, will take place, provided 1. — that the level in A is higher than in B; 2. — the levels are open to atmospheric pressure and 3. — the connecting pipe is completely filled with water.

are of the same height. N.B. Flow will take place even if parts of the connecting pipe are above the levels in the containers (FIG. 8.4) provided that the tube is completely filled and *both levels are exposed to atmospheric pressure.* In such cases where the upper surface of the container must be closed a tube must connect the upper surface with the atmosphere. This is done in cases of blood transfusion, in which case the container must be closed hermetically to prevent contamination. (FIG. 8.5).

Application of hydrostatic pressure. The principles of flow caused by hydrostatic pressure differences are applied in medicine in the emptying and filling of body cavities. This way blood or other fluids can be given intravascularly, blood can be taken from the patient and abscesses can be washed out.

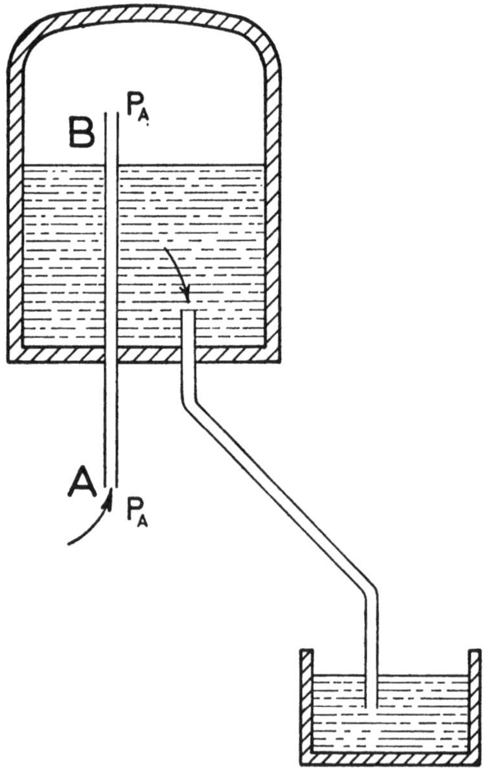

FIG. 8.5 In the case of closed containers the upper surfaces are exposed to atmospheric pressure by means of a pipe (AB) which is inserted from under the containter. PA = atmospheric pressure.

Measurement of pressures

(*a*) *Atmospheric pressure.* Toricelli was the first person to measure atmospheric pressure. He filled a tube about 900 mm long and closed at one end with mercury. He held the other end closed and submerged it under mercury standing in a low container (FIG. 8.6). He then opened the end under the mercury in the container, and observed that the length of the column of mercury in the tube dropped to 760 mm. In the experiment the pressure (P_1) on the surface of the open container is equal to the atmospheric pressure, while the pressure (P_2) at the bottom end of the long tube is caused

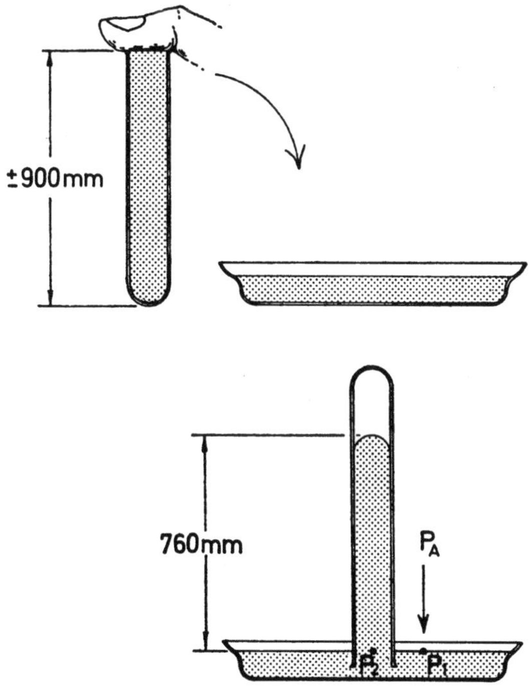

FIG. 8.6 Schematic representation of the method of Toricelli for the measurement of atmospheric pressure.

by the weight of the mercury column. Because the system is in equilibrium, i.e., no movement takes place, the two pressures must be the same, i.e., $P_1 = P_2$. But $P = \rho gh$, and therefore $P_A = \rho gh$, where ρ = density of mercury, h = height of mercury column and g = the gravitational constant.

The height of the mercury column is therefore proportional to the atmospheric pressure and thus provides an indication of the value of atmospheric pressure. Toricelli's apparatus is called a *barometer* and the height of the mercury column is called the barometric height. The normal barometric height at sea level and at a temperature of 0°C is 760 millimetres of mercury. The pressure at sea level at 0°C can also be expressed as 1 Atmosphere (1 ATM), which in SI units is 101 000 pascal or 101 kilo-pascal*. The atmospheric pressure changes from day to day and from place to place. The higher one is above sea level, the less is the air pressure, because the mass of air causing the pressure becomes smaller with increase

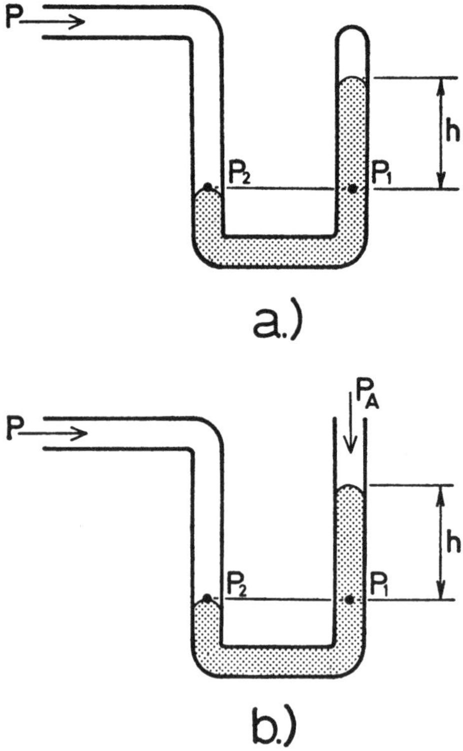

FIG. 8.7 Schematic representation of (*a*) a closed and (*b*) an open manometer.

*ρ mercury = 13,6 × 10³ kg/m³, g = 9,77 Pa m³/kg, h = 0,76 m. Pressure = ρgh = 13,6 × 10³ × 9,77 × 0,76 = 101 kPa

in height. Normal atmospheric pressure is very often referred to as zero pressure, and other pressures are expressed as *relative pressures*. If the pressure is greater than atmospheric pressure one speaks of a *positive* pressure and if it is less than atmospheric pressure it is called *negative* pressure.

(*b*) *Fluid pressure* is generally measured by means of *manometers*. There are two types of manometers, namely closed and open types (FIG. 8.7). Both types consist of a U-shaped tube containing mercury or water. One end of the U-tube is connected to the fluid, the pressure (P) of which is to be measured.

Closed manometers. In closed manometers one end of the U-tube is closed. There is a vacuum between the mercury or water and the closed end. When the system is in equilibrium the pressures in the two arms of the U-tube are the same, i.e., $P_1 = P_2$. In this case P_2 is the pressure which is being measured, and P_1 the pressure caused by the mercury column. As mentioned earlier the pressure exerted by the mercury column can be given by $P_1 = \rho gh$. A closed manometer therefore works precisely on the same principles as a barometer. A barometer is actually a simplified form of a closed manometer used to measure atmospheric pressure. A closed manometer measures the absolute pressure exerted by a gas or liquid.

Open manometers. In open manometers the pressure in the open tube is caused by the atmospheric pressure plus the pressure of the mercury column, i.e. $P_1 = P_A + \rho gh$. This means that the length of the mercury column indicates the difference between the atmosphere pressure and the unknown pressure, or in other words, it measures relative pressure. This relationship can be expressed as $\rho gh = P_1 - P_A$.

Fluids used in manometers. As mentioned earlier, water or mercury can be used in manometers. If the pressure that is to be measured is high, mercury is used. The density of mercury is 13,6 times that of water, and a mercury column will therefore indicate the same pressure as a water column 13,6 times its length. When the pressure to be measured is low and variable, water should be used instead of mercury.

(*c*) *Blood pressure*. Blood flow is caused by pressure differences. Two methods are used to measure blood pressure, namely direct and indirect methods. In the *direct method* one end of a sterile plastic tube is connected to the manometer, while the other is connected to a hollow needle. To measure the blood pressure the sharp point of the needle is inserted into a blood vessel. In the *indirect method* a special rubber bag, provided with two tubes, and surrounded by a linen bag, is used. The rubber bag is wound round the arm of the patient and inflated through one of the tubes until the air pressure in it completely stops blood flow in the main artery of the arm. The other tube is connected to a manometer and measures the pressure in the bag. Air is slowly let out of the bag, while the artery is auscultated by means of a stethoscope, held against the arm over the artery. When the pressure in the bag falls below the highest limit of the blood pressure in the artery, intermittent turbulent flow takes place and it can be heard by means of the stethoscope. The pressure (systolic pressure) is then read off the manometer. When the pressure in the bag equals the lower limit of the blood pressure, the flow becomes laminar, the sound disappears and the pressure is read off again (diastolic pressure). This apparatus is known as the sphygmomanometer, and is used extensively in medical practice. Typical values of systolic and diastolic pressures are 16 and 11 kPa (120 and 80 mm Hg), respectively.

(*d*) *Pressure transducers* can also measure pressure directly by converting it to an electrical signal. The electrical signals can be registered on an input apparatus (Chapter 7). This method is more sensitive and reliable than ordinary manometers, but unpracticable for everyday use in clinics. Consequently pressure transducers are used mainly in experimental laboratory work.

CHAPTER 9

Gases and Respiration

Introduction

Gas is a state of matter. The atmosphere consists of a mixture of gases, one of these being *oxygen*. Oxygen is absolutely essential for life, so much so that without it, life is only possible for a few minutes. Another gas, *carbon dioxide*, is also vital. This gas is one of the end products of chemical reactions which are continuously taking place in the human body. Because of their importance it is proper that we should look briefly at the laws which deal with the behaviour of gases under different circumstances.

The *lungs* are organs especially geared to provide oxygen (O_2) for the body and to clear carbon dioxide (CO_2) from it. Oxygen is taken up from the air and the excess carbon dioxide is expired. The tube system, transporting the gas mixture to and from the lungs, together with those parts of the lungs where gas exchange takes place and also the nerves which control the system, constitute the respiratory system. If the respiratory system does not function properly, it is still possible to provide sufficient oxygen and to remove the carbon dioxide with the aid of artificial *respirators*.

Pressures caused by intermolecular forces

The molecules of a gas are in constant movement. As a result of this, they collide with each other and if they are in a container, with the walls of this as well. The collisions against the walls of the container cause a pressure exerted on them. Assume that a certain number of gas molecules is present in an air-tight container, and that the volume of the container is V (Fig. 9.1).

The pressure, P, exerted by the molecules, depends on their intermolecular forces, their movement, and their collisions with the walls of the container. If the temperature is kept constant and the gas is compressed by reducing the volume of the container, the molecules come closer to each other and the intermolecular forces increase. This results in an increased pressure. In Chapter 1 it was shown that an inverted proportionality exists between pressure and volume (FIG. 1.5), i.e.

$$P = \frac{k_1}{V}$$

where k_1 represents a constant of proportionality.

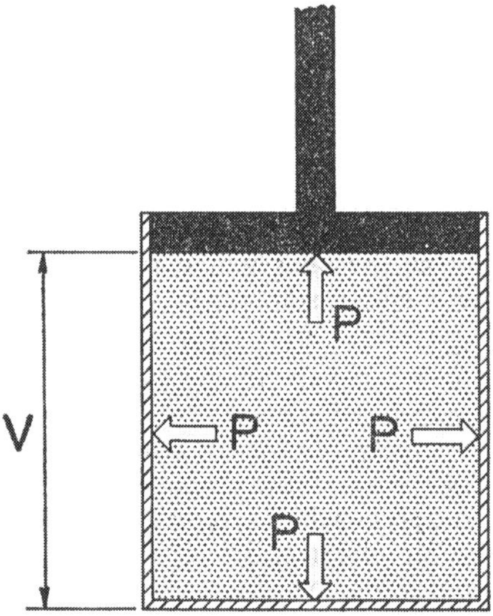

FIG. 9.1 The molecules of a gas exert a pressure (P) in all directions on the walls of the container. V is the symbol for the volume of the container.

A further property of gases is that when warmed, their pressure increases proportionally with the absolute temperature (T in kelvin), provided that the volume of the container and the total number of molecules in the container remain unchanged, i.e.,

$$P = k_2T$$

where k_2 is the constant of proportionality in this case.
The reason for the increase of pressure is the increased activity of the molecules when energy in the form of heat is supplied to them.
If the temperature and the volume are kept constant but the number of molecules change, the pressure is then proportional to the number of molecules, i.e.:

$$P = k_3n$$

In Chapter 1 it was shown that the three equations can be combined to form a single equation:

$$P = (k_1 k_2 k_3) \frac{Tn}{V}$$

$k_1 \times k_2 \times k_3 = R$, and it is called the *universal gas constant.* The equation can thus be written as:

$$P = \frac{nRT}{V} \text{ or } PV = nRT$$

The two pressures that have so far been discussed, namely the pressure due to the gravitational force and the pressure caused by the molecular movement both operate in the atmosphere. The pressure caused by the weight of air compresses the bottom layers of the atmosphere. This compression causes an increased pressure, called the internal pressure. The two pressures balance each other, and this is the reason why air movement does not take place under normal circumstances. The relationship which exists under normal circumstances can be expressed by the formula:

$$P_A = \frac{nRT}{V}$$

or

$$P_A = \frac{n}{V} RT$$

P_A = is the atmospheric pressure

This means that at normal atmospheric pressure and constant temperature there is a certain number of molecules per unit volume in a gas, or in other words, it is the concentration of molecules that determines the pressure. This means that the pressure in a container can be reduced either by the removal of gas molecules, or by increasing the volume of the container. In both cases the concentration of the molecules becomes reduced and the pressure becomes smaller. If the pressure becomes less than the atmospheric pressure one speaks of sub-atmospheric or negative pressure. If all* the molecules are removed from the container, one speaks of a *vacuum.* If the concentration of the molecules in the container is increased, by introducing more gas, or by reducing the volume of the container, the pressure is increased above atmospheric pressure and one speaks of a positive pressure.

Principles of pulmonary function

Pulmonary function is based on the laws mentioned in previous chapters and in the foregoing text. The lungs are enclosed in an

*It is practically impossible to remove all molecules of a gas from the container.

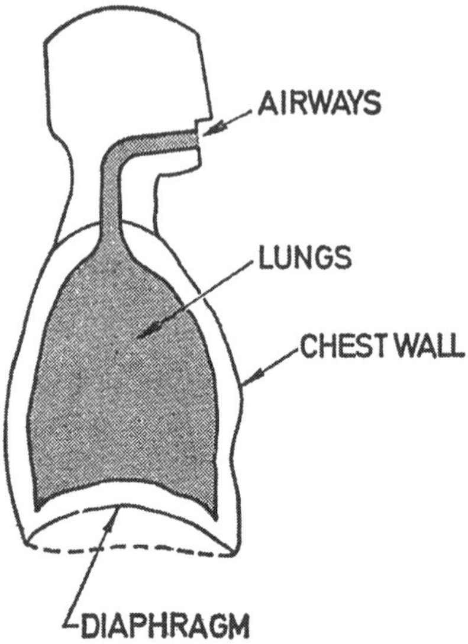

FIG. 9.2 Schematic representation of the chest cavity, air passages and lungs.

air-tight container, the chest cavity. The ribs, the sternum, the spine and the skeletal muscles joined to the bony structures form the walls of the chest cavity, while at the base it is closed by the dome-shaped diaphragm (FIG. 9.2).

Contraction and relaxation of the diaphragm and other muscles joined to the ribs and other bones change the volume of the chest cavity. The lungs themselves are elastic sac-like structures which can therefore expand and relax. The lungs are enclosed in a double layer membranous covering, the pleura. The outside layer of the pleura is joined to the inner walls of the chest cavity, and the inside layer of the pleura is joined to the outer surface of the lungs. In the space between the two layers of the pleura (intrapleural space), there is a small amount of fluid (intrapleural fluid). This fluid layer retains the two layers of the pleura against each other, but permits them to slide over one another.

The lungs actually consist of millions of air sacs (alveoli) that are connected to the main air passages, by small tubular structures (bronchioles). The smaller air passages (bronchi) join each other to form the trachea thus establishing contact between the alveoli and the atmosphere. The multitude of alveoli together constitute the intrapulmonary volume.

The dome-shaped diaphragm as well as other respiratory muscles, are under the control of nerves that conduct impulses from the brain to these muscles. With inspiration, the diaphragm contracts, becoming straighter. At the same time the other muscles also contract. The net result is an increase in the volume of the chest cavity. The outer pleural layer moves with the chest wall and pulls the inner layer out along with the lungs. As a result of this the volume of the lungs increases. Because the total number of molecules and the temperature do not change, the pressure in the alveoli will fall in inverse proportion to the increase in volume. This means that the pressure in the lungs is negative at this moment, and the atmospheric pressure is therefore higher than the pressure in the lungs. As a result of this pressure difference, air will flow from the atmosphere into the lungs. This increases the number of molecules in the lungs and the pressure increases. When the atmospheric and intrapulmonary pressures are equal, the flow will stop.

With expiration precisely the opposite occurs. The diaphragm and the other muscles relax and consequently the volume of the chest cavity and the lungs becomes smaller. Because of this decrease in volume, the concentration of gas molecules increases and therefore the pressure in the alveoli also increases. When the pressure in the alveoli becomes higher than the atmospheric pressure, air will flow from the lungs to the atmosphere.

Partial gas pressures and the exchange of gases between the lungs and the blood

In a mixture of gases each gas contributes towards the total pressure exerted by the mixture. This contribution is called the partial pressure of the gas and, is in direct proportion to the percentage of the gas in the mixture. Air is a mixture of gases, consisting mainly of oxygen (21%) and nitrogen (79%). At a temperature of 0°C the atmospheric pressure is 100 kPa at sea level. This means that oxygen contributes 21% or 21 kPa to the total air pressure and nitrogen 79% or 79 kPa. The concentration of carbon dioxide in the atmosphere is so small in comparison with oxygen and nitrogen, that it hardly contributes towards the atmospheric pressure.

The partial pressures of O_2, N_2 and CO_2 are indicated by the symbols P_{O_2}, P_{N_2} and P_{CO_2}, respectively. Gases dissolve to a certain extent in liquids when they come into contact with them and create a partial pressure in the liquid. If a difference in partial pressure between a gas in the gas mixture and the gas in the liquid exists, it causes movement of the gas between the gas mixture and the liquid. If the partial pressure of a gas in the gas mixture is the same as in the liquid, no movement of gas will take place and there is equilibrium.

When blood flows through the blood vessels of the lungs, it comes into close contact with the air in the alveoli. Blood and air are actually only separated by two thin membranes. Because the Po_2 in the alveoli is greater than the Po_2 in the blood, oxygen diffuses from the alveoli into the blood. Diffusion proceeds until the partial pressures in the two compartments are equal. After the oxygen has dissolved in the blood, most of it (98,5%) immediately diffuses through the membrane of the red blood cells and combines with the haemoglobin in the cells, only a small percentage remaining in the dissolved form. It is actually this small percentage of oxygen, and not the oxygen which is combined with haemoglobin, which produces the considerable partial pressure of 13 kPa. This pressure is much higher than the partial pressure of oxygen in the tissues.

When the oxygen-rich blood arrives at the tissues it gives up its oxygen, again as the result of difference in partial pressures. As the Po_2 in the plasma falls, the haemoglobin releases its oxygen and thus more oxygen is made available to the tissues. When the venous blood reaches the lungs, the Po_2 in the lungs is higher than in the blood, and oxygen again diffuses into the blood. Nitrogen only dissolves to a very slight extent in blood and is therefore not transported in the physiological sense. It is of no importance to us since it does not bind with haemoglobin and is not required by the tissue cells.

During metabolism the tissues produce, amongst other substances, considerable amounts of carbon dioxide, and therefore the partial pressure of carbon dioxide in the tissues is relatively high. The blood flowing through the tissues has little dissolved carbon dioxide in it, and therefore the Pco_2 in the arterial blood is low. This causes carbon dioxide to diffuse from the high Pco_2 in the tissues to the lower Pco_2 in the arterial blood. The venous blood that arrives in the lungs contains a high concentration of carbon dioxide and the Pco_2 is therefore considerable. Because the Pco_2 in the venous blood is higher than in the alveoli, the carbon dioxide diffuses from the blood to the lungs. From the lungs it passes together with air through the air passages to the outside.

Normally the Po_2 in the atmosphere is 21 kPa at sea level, producing a Po_2 in the alveoli of about 13,5 kPa. Even if this Po_2 falls to 11kPa the haemoglobin in the blood is saturated with oxygen. If, however, the Po_2 decreases below this level the haemoglobin is not saturated any more, and the tissues receive an insufficient amount of oxygen. This condition pertains at high altitudes where the atmospheric pressure and therefore the Po_2 is considerably lower.

The condition can be alleviated by artificially increasing the oxygen content of the inspired air, using oxygen cylinders. Also in certain diseases, such as heart failure, certain types of anaemia and certain types of lung disease (pneumonia and bronchial asthma), the inspiration of air that contains more than 21% oxygen, i.e., with a Po_2 of more than 21 kPa, can be life-saving.

Artificial respirators

Respiration is controlled by nerve impulses from the brain. The impulses are produced in the respiratory centres of the brain and are transmitted to the respiratory muscles. Generally respiration is a subconscious and involuntary process, but it can also be altered voluntarily. An adult person breathes about 14-18 times a minute. The whole mechanical part of respiration, i.e., from the production of impulses in the brain to the increase of chest cavity and lung volume, serves to produce a pressure difference between the atmosphere and the lungs, in order to cause a flow of air. The amount of air that flows is dependent on (1) the pressure difference and (2) the resistance that the air passages offer towards the flow. Various conditions, such as a fault in the respiratory centres of the brain where the impulses are produced, or in the nerves that conduct the impulses, or in the respiratory muscles which regulate the lung volume, result in failure to produce the required pressure differences, thereby reducing the air flow, sometimes even completely. Even if the above-mentioned mechanisms function normally, the flow can be reduced through damaged or constricted air passages. If the flow to the lungs is sufficient, the body can nevertheless experience an oxygen deficit if the diffusion of oxygen from the alveoli to the blood is not optimal. Such a condition exists in lung oedema, or when the blood circulation through the lungs is reduced.

Respirators can be used to correct faults in respiration. The purpose of respirators is to produce a sufficient pressure difference between the two ends of the air passages or to increase an existing but insufficient pressure difference. This can be done in two ways:

(1) By producing a pressure in the lungs that is less than the atmospheric pressure, just as it is in the case of normal respiration. Such a pressure is a negative pressure and respirators that operate on this principle are called negative pressure respirators.

(2) By producing pressure at the mouth that is greater than atmospheric pressure. In this case one speaks of positive pressure respirators.

Negative pressure respirators reduce the pressure in the lungs, i.e., they imitate normal respiration An example of a negative respirator is the so-called "iron lung" (FIG. 9.3).

The patient is placed in a metal case, in such a way that only his head remains outside of it. The pressure in the metal case (P_1) can be regulated. When this pressure decreases, the chest expands, and

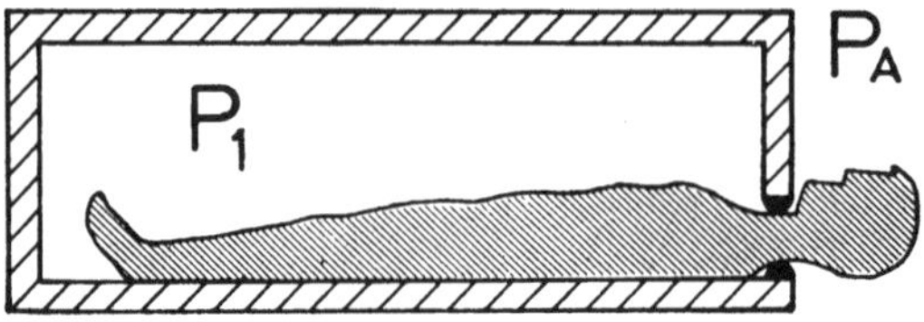

FIG. 9.3 Schematic representation of an "iron lung". P_A = atmospheric pressure, P_1 = pressure inside the apparatus.

thereafter inspiration proceeds normally. For expiration the pressure P_1 is increased. Smaller apparatuses, that only enclose the chest of the patient, are also used. The principles involved are the same as those for the "iron lung". The advantage of the small apparatus is that it is portable. Negative pressure apparatuses are used very rarely, because of the problems encountered with them. The iron lung is too heavy, while it is very difficult to get an airtight connection between the small apparatus and the patient.

Positive pressure respirators produce a high pressure at the entrance to the air passages. The high pressure is produced by compressing an elastic bag that has been filled with air (FIG. 9.4). Expiration takes place passively, as the expanded lungs resume their normal shape due to their elasticity. Direction of air flow is controlled by special valves located in the system.

In modern respirators, as shown in FIG. 9.4, the pressure changes and the opening and closing of these valves are controlled automatically. Respirators can be separated into two groups.

1. Automatic respirators in which the frequency of respiration is controlled automatically. These are used in patients who cannot breathe at all.

2. Supporting respirators which are used when the patient can breathe, but insufficiently. In this case his attempts at respiration are used to activate the respirators.

Both types of apparatus can further be divided into three subgroups, according to the manner in which the duration of inspiration is controlled.

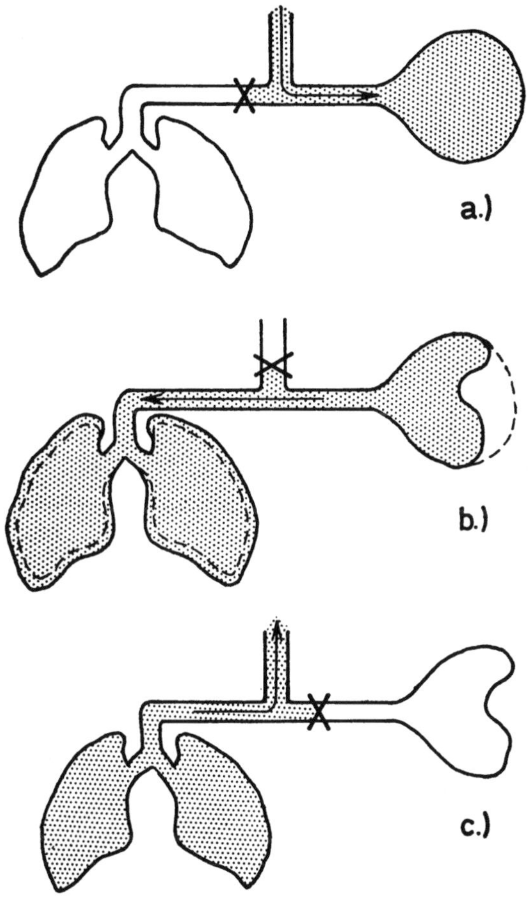

FIG. 9.4 The principles of positive pressure respirators. (*a*) The sac is filled with air, (*b*) it is compressed to force the air to the lungs, and (*c*) expiration takes place because the stretched elastic lungs resume their normal shape.

(*a*) Pressure-controlled respirators in which the pressure in the sac is increased slowly until a predetermined pressure is reached. The amount of flow is determined by the resistance in the air passages. A disadvantage of this apparatus is that a constriction of the air passages increases the resistance, with resultant decrease of inflow for the same pressure difference. In these cases the increase of pressure is fast, since the airflow out of the sac is slow.

(*b*) Volume-controlled respirators provide the patient with a definite volume of air, and therefore, the resistance of the air passages does not play a role. After a certain volume of air has left the apparatus to the lungs the apparatus changes over to expiration.

(*c*) Time-controlled respirators cause inspiration for a definite time, and then change over to expiration.

CHAPTER 10

Liquids and the Circulatory System

Introduction

Water is the compound which is present in the body in by far the greatest amounts, and without which life is impossible. The importance of water in the body can be seen from the fact that the percentage of water in an average adult male is between 54-74 per cent, and in an average adult female it is 45-60 per cent of the body mass. Water has the following functions in the body.

A. It serves as a solvent for the great variety of materials in the body. This is important, because most materials can only react with each other in solution.

B. It serves as a medium of transport for the dissolved nutrients, end products of metabolism, hormones, electrolytes, gases, etc., between the various organs and tissues.

C. Because of its high specific heat and high evaporation heat, water plays an important role in the regulation of body temperature (See Chapter 12).

In order to perform the above functions, water must continuously be kept moving. This is achieved by the circulation of blood in the blood vessels. The pumping action of the heart supplies the force necessary for the flow of blood, while the blood vessels serve as a distribution system taking the blood to the various parts of the body. Blood consists of cells and a fluid portion, the plasma. There are three types of cells in the blood, namely red cells, white cells and platelets. About 90% of the blood plasma is water, the remaining 10% consisting of all sorts of chemical substances that are dissolved or suspended in the water.

As already mentioned (Chapter 8), fluids flow as a result of pressure differences between the two extremities of a tube system. The relationship between the flow intensity and the pressure difference is described by the following equation:

$$I = \frac{P}{R}$$

I = flow intensity

P = pressure difference

R = resistance presented by the tube to the flow of the fluid.

The meaning of resistance has already been discussed (Chapter 8) and consequently only the production of pressure differences will be discussed here. For the sake of clarity the reader is advised to revise Chapter 4, because a good understanding of Ohm's Law makes for easier understanding of the functions of the circulatory system.
In order to produce movement a force is required. In the case of electricity the force is called the *electromotive force* (EMF). This force causes acceleration of electrons, but as a result of friction between the electrons and the metallic structure a counter-force develops which acts against the movement of electrons. The opposing force increases with the speed of the electrons and when it reaches the same magnitude as the EMF, the net force on the electrons becomes zero and therefore the electrons continue to move at a constant rate in a straight line. In the case of electricity the electromotive force is produced by a battery. In the battery are two different metal plates at which different chemical reactions produce a charge difference. When the charge difference tends to decrease, by, for example, the draining off of electrons, chemical reactions take place again and the charge difference is re-established.

"Hydromotive force"

In the circulatory system there is a force that coincides with the electromotive force of electricity. In electricity the metal plates are charged with electrons, and in the circulatory system the large blood vessels are charged (filled) with blood. The blood stretches the elastic blood vessels. As a result of the stretching the elastic vessels exert a force on the blood in the vessels. Since the blood vessels have greater elasticity at one end of the tube system than at the other, there is a pressure difference between the two ends. This pressure difference causes the blood flow. When the fluid is flowing in the tube system the pressure difference between the two ends decreases. The heart, however, replaces this blood by its pump action.
By analogy with the electromotive force that causes the flow of electricity, one may call the force causing the flow of liquid a *hydromotive force* and similarly that which causes blood flow a *haemomotive force*. We shall use the abbreviation HMF in both cases.
Just as in the case of electricity, there is an opposing force that develops as a result of friction. In this case it develops between the fluid layers of the blood. When the HMF and the opposing force are equal, the flow proceeds at a constant rate. The pressure difference between the two ends of the tube system indicates the magnitude of both of the forces.
The structure of the circulatory system has already been discussed (Chapter 8). The system consists of two parts, namely the general

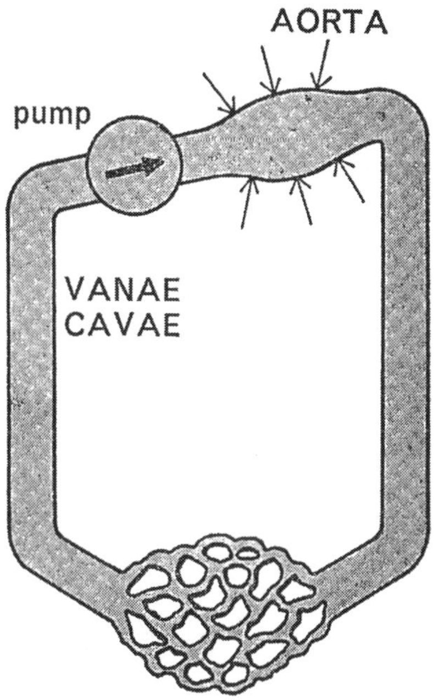

FIG. 10.1 Schematic representation of the stretching of the aorta by the blood and the force exerted by the elastic tissue on the blood.

body circulation and the lung circulation. The principles according to which the two parts function are the same, and therefore only the body circulation, the so-called systemic circulation, will be discussed. Blood reaches the system from the left heart through the aorta and returns to the heart through the venae cavae. In the circulatory system the gravitational force plays a minor part in the production of the HMF.

As mentioned, it is the elastic force that provides the HMF. When elastic material is stretched, a force is created between the molecules of the material. This force tends to bring the material back to its normal shape and length. A practical example is the stretching of an elastic band. The greater the stretching, the greater the opposing force that is being produced. The opposing force is also determined by the type of elastic material. A thin soft elastic band stretches easily, i.e., the opposing force is small. More force is needed to stretch a thicker, stronger elastic band, i.e., the opposing force that is produced is greater.

The blood vessels contain elastic tissue in their walls. The amount of elastic tissue present in the vessel walls varies in the different parts of the circulatory system. It is considerable in the aorta, while only a small amount is found in the venae cavae. This means that the aorta is considerably more elastic than the venae cavae.
The hydromotive force that forces the blood to flow from the aorta to the venae cavae varies continuously (FIG. 10.2).

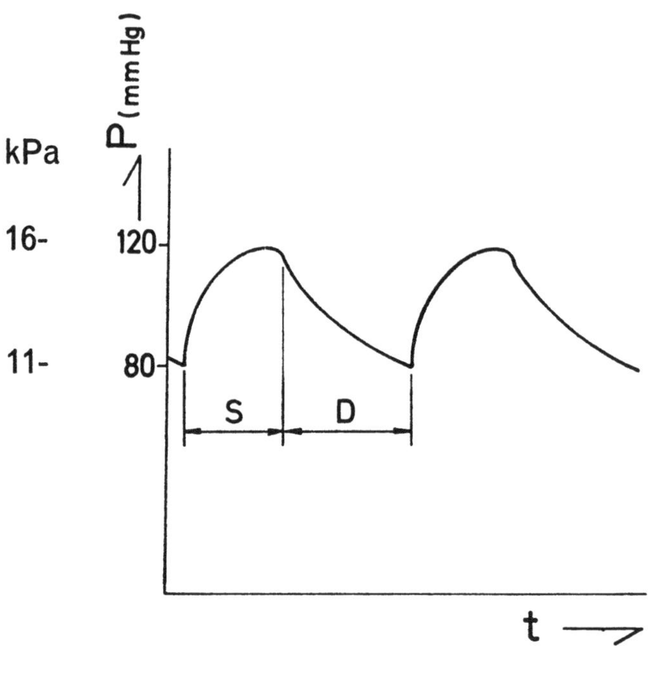

FIG. 10.2 Pressure variations in the aorta. P = pressure, t = time, S = systole, D = diastole.

The pressure in the aorta decreases as blood flows out of it through the circulatory system to the venae cavae. Before the pressure falls too low, the heart refills the aorta once more and the pressure rises again. The pressure in the aorta is therefore always higher than in the venae cavae thus ensuring continued circulation. The period during which the heart pumps blood into the aorta is called *systole* (S) and the period during which the heart rests is called *diastole* (D). The pressures at the end of the two periods are called the systolic and diastolic pressure, respectively. Mean values of the two pressures are

16 and 11 kPa, respectively. Since the blood pressure plays such an important role in the circulation, it is essential that we know the blood pressure of a patient. Methods of measurement of the blood pressure have already been described (Chapter 8).

The heart as a pump

Fluid generally flows from a point with a high pressure to a point where the pressure is lower. By definition, a pump is an

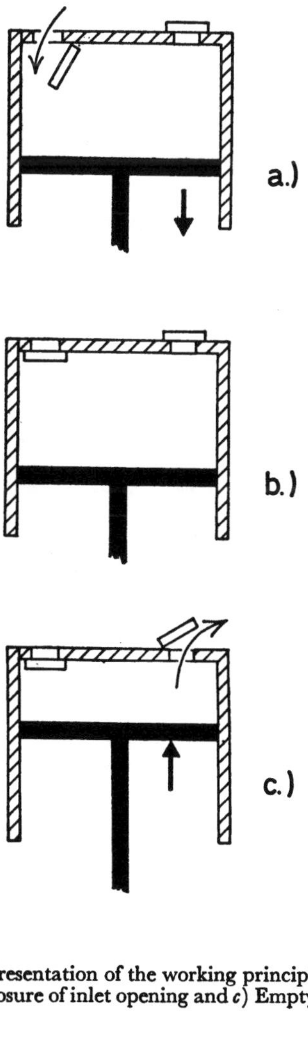

FIG. 10.3 Schematic representation of the working principles of a pump. *a*) Filling, *b*) Closure of inlet opening and *c*) Emptying.

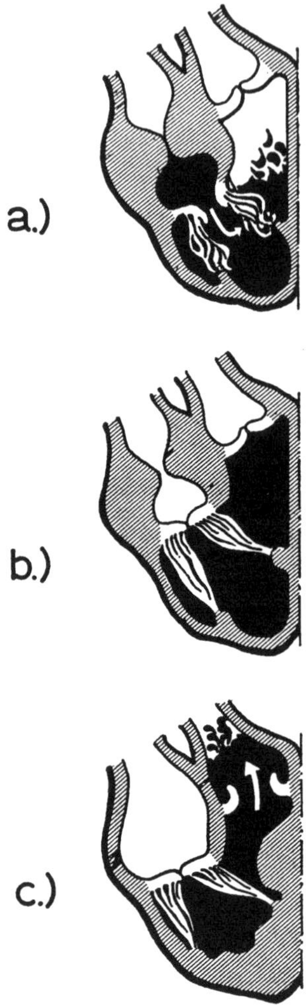

FIG. 10.4 Schematic representation of the heart as a pump. The three phases are as described in Fig. 10.3.

apparatus that forces the fluid from a lower to a higher pressure. Most pumps work according to a simple principle, namely, fluid enters the pump through an opening (filling phase). This opening closes and pressure is exerted on the fluid in order to force it through another opening (pumping phase) (FIG. 10.3). The force causing the requisite pressure is obtained from a mechanical movement. A very simple example of such a pump is the bicycle pump.

The heart consists of four chambers, namely, the two upper chambers or atria, and the two lower chambers, the ventricles. In FIG. 10.4 only one side of the heart is shown, because the two halves function precisely according to the same principles. It is actually the ventricles which perform the pumping function. The wall of these chambers contains muscle tissue that contracts and relaxes regularly about once every second. When the muscle contracts the input openings to the chamber close, while the pressure caused by the contraction increases, and forces the output openings open, causing the blood to flow into the arteries. The chambers of the heart and the blood vessels form a closed system of tubes, and as a result of this orientation, the blood pumped by the heart into the arteries returns to the heart by the veins after it has circulated through the tissues. Blood fulfills its function as it circulates through the capillaries, exchanging nutrients, metabolic breakdown products and gases with the tissue fluids.

In discussing blood circulation through the heart one may start at any point in the system (FIG. 5.12). We shall start at the right atrium.

After blood has circulated through the systemic vessels, it returns to the right atrium of the heart through the two venae cavae. From the right atrium it flows through the atrio-ventricular opening to the right ventricle. When the ventricles contract, the valves between the atria and ventricles close, and the increased pressure in the ventricle opens the valve in the opening between the right ventricle and the pulmonary artery in order to force blood through it. As the blood circulates through the lungs it sheds its excess carbon dioxide and takes up oxygen. The oxygen saturated blood returns to the left atrium through the pulmonary veins. From the left atrium the blood flows through the atrio-ventricular opening into the left ventricle. When the left ventricle contracts the valves between the left atrium and the left ventricle close, while the valve between the ventricle and the aorta is forced open. As a result blood flows into the aorta and is distributed to the tissues. The two atria are actually very weak pumps that only help to complete the filling of the ventricles. The ventricles are strong pumps which force blood against the resistance of the blood vessels to the lungs and to the rest of the body. It must be stressed that the two atria contract together, followed after about an eighth of a second by the simultaneous contraction of the two ventricles.

The chambers of the heart function according to the same principles as ordinary pumps. During the filling phase the blood enters the ventricle from the atrium through the atrio-ventricular opening. When the ventricles are full and begin to contract, the atrio-ventricular valves close, and the outflow valves open and blood is pumped into the corresponding arteries. It can therefore be said that the heart forces blood from the veins where the pressure is low, to the aorta where the pressure is higher, thus creating a pressure difference that causes blood flow throughout the body.

Heart-lung machine. During open heart operations the surgeon is working on the heart and it cannot perform its normal function as a pump. An artificial pump must be used to keep the blood circulating. It is clear from the above description that the heart actually consists of two pumps, right and left, and therefore it must be replaced by two pumps, namely one that pumps blood through the body and one that pumps it to the lungs. This necessitates many connections between the tubes of the machine and the blood vessels, causing many problems, and it is therefore much easier to use one pump only and to replace the lungs with an artificial lung. The artificial lung ventilates the blood, i.e., it causes it to take up oxygen and release carbon dioxide, while the artificial pump forces blood into the aorta. Blood circulates through the body, and then returns to the artificial lung via the two venae cavae. This so called heart-lung machine enables the surgeon to proceed with the operation, knowing that the various organs of the body are assured of an adequate blood circulation.

During the operation the pumping activity of the heart can be stopped by *fibrillating* the heart. Fibrillation can be caused by cooling or by the use of an electronic fibrillator. After the operation the heart can be reconnected to the circulatory system and its normal pumping function is resumed after *defibrillation.*

It must be stressed that the heart-lung-machine can only be used for a few hours, the reason for this being that the apparatus is constructed of metal and plastic which may cause the destruction of red blood cells and even coagulation of the blood.

Heart valves

The *one-way* flow of blood through the heart is ensured by the heart valves. Heart valves are present at the input and output openings of the ventricles. The valves can only open one way and are regulated by pressure differences. The actual construction of the heart valves will not be described here but the working of the normal valve between the ventricle and the aorta will be explained by means of a simple model (FIG. 10.5).

Let us assume that the pressure in the ventricle is P_V and the pressure in the aorta is P_A. The valve can rotate about an axis and can only open in the direction shown by the arrow, because the frame stops it from opening in the other direction. When the pressure in the ventricle (P_V) exceeds the pressure in the aorta (P_A) the valve is forced open in the direction of blood flow. If the pressure in the

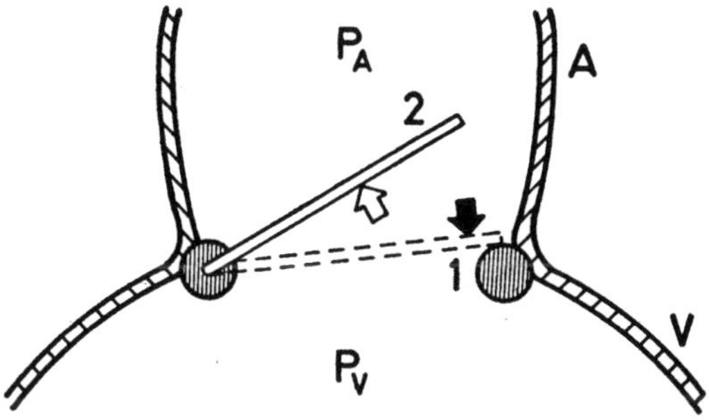

FIG. 10.5 Schematic representation of a valve between the ventricle (V) and the aorta (A). P_A and P_V are the pressures in the aorta and in the ventricle, respectively. Positions 1 and 2 represent the closed and open positions of the valve, respectively.

aorta is greater than in the ventricle, the valve is forced against the frame and blood will thus not flow back to the ventricle.

The valves of an intact heart work on the same principles as the model described above. Sometimes the opening of the valve becomes constricted, a condition called stenosis. In another condition called valvular incompetence, the valve opening is malformed to such an extent that the valve does not close it properly. Should the above conditions become advanced, heart function becomes affected to such an extent that the valves must be replaced by artificial valves.

The principles according to which most artificial valves work are illustrated in FIG. 10.6. The valve consists of a light ball that can move in a metal framework. The ball can assume two different positions, and its position at any time is determined by the direction of the net pressure on the ball. When the pressure in the chamber behind the ball is higher, it is forced into the forward position and this allows blood flow to take place. When the pressure is greater in the chamber in front of the ball, the latter is forced into the backward position and the opening is closed off, thus preventing back flow.

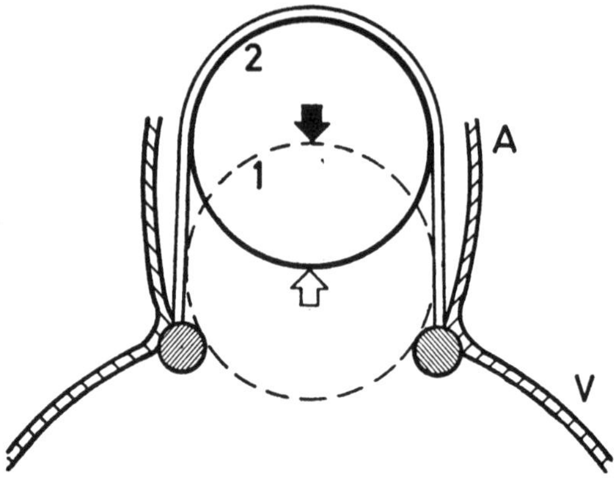

FIG. 10.6 An artificial valve between the ventricle (V) and the aorta (A). Positions 1 and 2 are the closed and open positions, respectively.

CHAPTER 11

Fluid Mixtures

Introduction

When two substances are equally distributed in a container or in a space one speaks of a mixture. If the substance which has the higher concentration happens to be liquid, one speaks of a fluid mixture. The other substance in the fluid mixture can be in any of the other physical states, i.e., it can be a gas, a liquid or a solid. A knowledge of fluid mixtures is important, since the living body is actually also a fluid mixture. It is in fact a compound mixture, consisting of solutions and other mixtures.

In the case of fluid mixtures the components of the mixture can move in various ways and independently of each other. The different types of movement that can take place are *diffusion*, *osmosis*, *filtration* and *active transport*. Each of these processes is of utmost importance for the normal functioning of the body. For example, the fluid exchange between the tissues and the blood is controlled by filtration and osmosis, while the kidneys use all four processes to perform their functions. The design and functioning of the artificial kidney are based on a combination of the same physical processes.

Solutions

A substance that is mixed with a liquid can behave in two ways, namely: 1. it can break down into molecules, or 2. it can retain the form it had before mixing.

The process by which the substance is broken down to its molecular level is called the *solution process*, and the mixture obtained is a *solution*. Substances that dissolve are called *soluble* substances. After the substance has dissolved, it is called the *solute*, and the liquid in which it has been dissolved is the *solvent*. The molecules of the solute spread out equally in the spaces between the molecules of the solvent. When the available spaces between the molecules are filled, the solvent cannot dissolve any more material and it becomes saturated.

The solubility of different substances in the same solvent differs. Similarly the solubility of a certain substance changes from solvent to solvent. Not only are solids soluble in appropriate solvents but so are gases and liquids. Water is the solvent encountered in tissue fluids and cells. Many chemical substances and the two gases, oxygen and carbon dioxide, dissolve in water. The amount of gas that dissolves is dependent on the partial pressure (Chapter 9) of the gas.

Other mixtures

If the substance that has been mixed with the liquid is insoluble, it remains in its initial form. According to the physical state of the substance concerned, one speaks of a suspension or of an emulsion. In the case of suspension, the substance is in the solid state, and in the case of emulsions, it is liquid. The distribution of the substance is even in both cases, just as in the case of solutions.

Suspensions can be divided into two groups, according to the size of the particles of the substances.

1. If the particles of the substance are so large that they sediment with the passage of time, the suspension is a normal one.
2. If the particles are so small that they do not sediment with time, one talks of a *colloidal* suspension. Substances that give colloidal suspensions are called colloids. Plasma proteins are examples of colloids.

In *emulsions* the liquid is present in the form of droplets. Emulsions can be temporary or permanent. An example of a temporary emulsion is oil mixed with water. After a time the oil droplets collect to form a layer on top of the water. An example of a permanent emulsion is milk.

Blood is a compound mixture consisting of solutions, suspensions and emulsions. It is a normal suspension of blood cells, a colloidal suspension of plasma proteins, a solution of electrolytes and an emulsion of fats.

Diffusion

In order to obtain even distribution of particles in solutions and colloidal suspensions, mixing takes place automatically under the influence of intermolecular forces. A solution or a colloidal suspension is actually not in equilibrium until concentration differences exist. The process of equalising is called diffusion and it can be explained by the presence of a diffusion force (Chapters 3 and 5). The mechanism is as follows. In all materials forces exist between molecules depending on the intermolecular distances concerned (Chapter 8). In regions where the concentration of the solute is great, the intermolecular forces are greater than in regions of low concentration. The force differences cause the molecules, or colloidal particles to move away from the area of high concentration, until concentration differences have been evened out. The diffusion of most nutrients from the alimentary canal to the blood, and from the blood to the tissues, takes place in this way. This also applies to the diffusion of oxygen from the lungs to the blood and from there to the tissues, and for the diffusion of carbon dioxide in the opposite direction.

Osmosis

Osmosis takes place in a solution when two regions between which concentration differences exist, are divided by a membrane that allows the movement of the solvent but not of the solute. Such a membrane is called a *semi-permeable membrane.* A semi-permeable membrane, like the membrane of cells (Chapter 5) has openings that allow small particles to pass through, but prevent large particles from passing through. In the case of osmosis the molecules of the solvent move from the area where the concentration of the solute is lower to the area where the concentration of the solute is higher. The concentration differences of the solute thus cause movement of the solvent in an effort to balance this concentration difference. This means that the solute exerts a suction force on the solvent, and this suction force is called the *osmotic pressure.*

Materials that dissolve in water, form crystals when the water becomes saturated with them. For this reason these materials are called crystalloids. Examples of crystalloids which occur in the body are Na^+, Cl^-, K^+ and Ca^{++}. The osmotic pressure caused by crystalloids is called the crystalloid osmotic pressure. When the material causing osmotic pressure is a colloid, one speaks of colloid osmotic pressure.

The particles of crystalloids are small, and therefore the pores that limit their movement are also small. It has been shown (Chapter 5) that the cell membrane has pores that are impermeable to Na^+. A difference in Na^+ concentration between the two sides of the membrane will cause movement of water through the pores. Under normal circumstances the extracellular fluid contains 0,9 per cent NaCl and at this concentration there is no osmosis. A 0,9 per cent sodium chloride solution is called an isotonic solution. If the concentration of NaCl outside the cells exceeds 0,9 per cent (hypertonic solution) water will move out of the cells (exosmosis). If the concentration of NaCl outside the cells is less than 0,9 per cent (hypotonic solution), water will move to the inside of the cells (endosmosis).

These principles of osmosis are of importance when one transfuses fluid into the blood vessels. The fluid must be isotonic with respect to the red blood cells, otherwise these cells will shrink (as a result of exosmosis), or they will swell (as a result of endosmosis). These changes influence the function of the red blood cells and as a result of swelling, they may even break up (haemolyse). The action of some purgatives, like Epsom Salts and Glauber salts, is based on the crystalloid osmotic pressure which they exert. A high concentration of these salts in the alimentary canal will cause an osmotic force that brings about the movement of water from the blood to the alimentary canal. This increased quantity of water excites peristaltic movements of the alimentary canal. For the same reasons the consumption of sea water is not thirst-quenching. Sea water, because it contains large amounts of dissolved salts, exerts an osmotic suction force on the water in the blood and sucks it into the alimentary canal. This means that the body actually loses water and the thirst becomes greater.

Filtration

Filtration is a process whereby certain components of a mixture are forced through a membrane, while other components are held back. Filtration is very often performed in the laboratory in order to separate materials. Different types of filters are used, one of which is filter paper. The flow of the liquid through the filter is regulated by hydrostatic forces. These forces were discussed in connection with the laws of flow (Chapter 8). Pressure differences therefore play a major role in filtration. Filtration can be regarded as the opposite of osmosis.

Exchange of fluids between the capillaries and the tissues

One of the functions of blood is the transport of oxygen and nutrients to the tissues, and the removal of carbon dioxide and metabolic breakdown products from the tissues. The exchange of these materials takes place in the capillaries, where the blood and the tissues are separated by a semi-permeable membrane. The pores of the membrane are so small that the membrane does not permit the colloid particles through, but only the crystalloid particles. The following processes are important in connection with the exchange of materials between the capillaries and the tissues.

1. *Diffusion.* Crystalloid materials and gases can diffuse through the pores. The blood contains large amounts of nutrients and oxygen when it reaches the tissues, i.e., the concentration of these substances in blood is high. On the other hand the concentration of these substances in the tissues is low since they are being continuously used up. Consequently there exist concentration differences between the blood and the tissues for the various substances, which therefore diffuse from the blood to the tissues.

The same holds for the metabolic breakdown products, the difference being that here the concentration of these substances is higher in the tissues than in the blood and consequently diffusion takes place from the tissues to the blood.

2. *Fluid exchange*. It has already been mentioned that blood in the blood vessels is under pressure caused by the overfilling of the vessels. This overfilling stretches the elastic vessels, which consequently exert a force (hydrostatic pressure) on the blood. This means that there is a pressure difference between the blood and the tissues. The fluid portion of the blood, together with the crystalloid substances, but not the blood cells and plasma proteins, will be forced through the small pores of the capillary membranes. In other words, the blood is filtered through the membrane. As blood circulates through the tissues and the water with the crystalloids leaves the vessels, the concentration of colloids in the blood increases and produces an osmotic pressure. There are therefore two pressures to consider in this regard, namely an outward hydrostatic pressure and an inward osmotic pressure. The outward hydrostatic pressure in the capillaries decreases progressively towards the venous end of the vessels. At the arterial end of the capillaries the outward pressure exceeds the inward pressure and at the venous end the inward pressure is the greater of the two. Under normal circumstances the fluid that leaves the vessels at the arterial end returns again at the venous end. The reason for this to and fro movement is to ensure continuous mixing of the fluid of the tissues with that of the blood and in this way supplying the tissues with nutrients and cleansing them of metabolic breakdown products The composition of the tissue fluids is regulated in this manner. If for one or other reason the net outward flow exceeds the net inward flow, excess water will move towards the tissues, this condition being known as oedema. Oedema will develop, for example, if the concentration of plasma proteins in the plasma decreases, causing a decrease in the colloid osmotic pressure. This will happen if too little protein is present in the diet (nutritional oedema), or if the kidneys excrete protein, as in nephritis or nephrosis. Oedema will also develop if the hydrostatic pressure in the capillaries rises above normal, as in the case of heart failure.

Active transport

Most membranes in the body show the ability to transfer certain substances from one side of the membrane to the other. Although the exact mechanism of this property of the membranes is still largely unknown, it can be shown experimentally that it does take place. The sodium pump (Chapter 5) is an example of this mechanism. Let us assume that a cell is placed in a solution of NaCl, as normally occurs in the human body. As explained before, the cytoplasm contains relatively little NaCl and the cell membrane is almost impermeable to Na^+. One would think, on the grounds of the preceding discussion, that Na^+ will diffuse from the outside of the cell to the inside as a result of the concentration differences and that the process will be slow because of the low permeability. If the cell were to be enclosed by a "dead" membrane, this is exactly what would happen. What in fact happens in a living cell is that Na^+ is removed from the cell to the outside as if it were acted upon by a force. This process is called active transport.

Another example of active transport is the process by which the thyroid gland takes up its iodine from the blood. Normally the thyroid cells have a high concentration of iodine which they use to form the important thyroid hormones. The blood that bathes the thyroid cells contains very little iodine. In spite of the high iodine concentration in the cells and the low iodine concentration in the blood, the thyroid cells are able to concentrate iodine. Active transport also plays an important role in the movement of various ions through the kidney tubule cells.

Principles of kidney function

Kidney function is based on the processes of filtration, osmosis and active transport. Both kidneys contain about one million kidney tubules. The arteries that supply blood to the kidneys divide many times to provide each tubule with a branch, and every branch comes into contact with the tubule twice (FIG. 11.1). At these contact

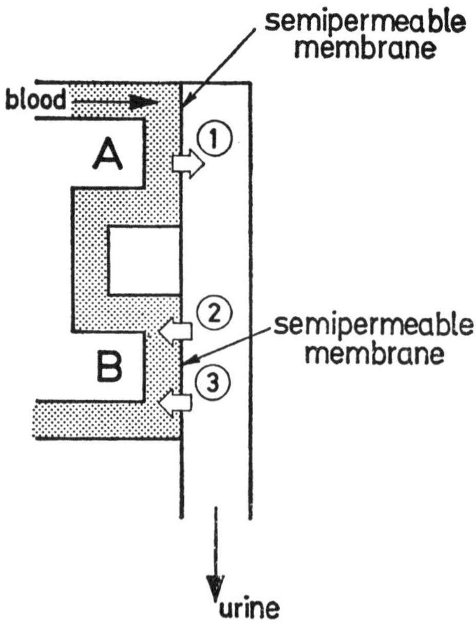

FIG. 11.1 Schematic representation of kidney function (see text).

points the blood vessel is separated from the urine tubule by semi-permeable membranes. The membranes at the two contact points differ from each other as follows. The first membrane (A on the diagram) has large pores that limit only the movement of blood cells and colloidal substances. Because the blood in this contact artery is under high hydrostatic pressure, the water with the crystalloid materials will be *filtered* through it into the urine tubule. The membrane at the second contact point (B on the diagram) allows only water and a few small molecules through, but has the capacity to transport substances actively. The process in the kidney tubules is called active *reabsorption*. Active reabsorption ensures that important substances that have passed into the proximal end of the tubule, are returned to the blood by the more distal part of the tubule and therefore not lost. One of the substances that returns in this way is Na^+, which normally cannot pass through the pores. The increase of Na^+ on one side of the membrane causes an *osmotic* pressure and a resultant suction of water back into the blood vessels.

Substances that stay in the urine tubules and leave the body together with urine, are:

(*a*) waste products;

(*b*) useful substances that are present in the body fluids in excess;

(*c*) a relatively small quantity of water in which the above substances are dissolved.

The main functions of the kidney are therefore the removal of toxic substances and maintaining the constancy of the body fluids.

Dialysis and the artificial kidney

Should the kidney become damaged its cleansing function can be performed artificially. The artificial regulation of blood composition is called haemodialysis or dialysis. In this process the blood is separated from a second fluid by a semi-permeable membrane. The principles of dialysis are the same as those of fluid exchange in the capillaries, except that the *dialysing fluid* (the second fluid) is continuously being changed. In this way it is ensured that the constitution of the dialysing fluid remains unchanged.

The semipermeable membrane allows all constituents of blood, except plasma proteins and blood cells, to pass through it. If the concentration of a substance in blood is greater than in the dialysing fluid the substance will move from the blood to the dialysing fluid. This is how waste products are removed from the blood. If the concentration of certain substances must be maintained at a constant level this concentration must also be the same in the dialysing fluid. On the other hand a substance will be transferred from the dialysing fluid to the blood if its concentration is higher in the blood provided that it can pass through the pores. In this way a substance like glucose can be given to a patient who cannot be fed normally. If, however, the substance cannot pass through the pores it will cause water to leave the blood vessels by means of osmosis.

In the artificial kidney the blood circulates in small channels that have semipermeable membranes for walls, on the outside of which circulates the dialysing fluid. To prevent coagulation of blood heparin is added to the blood before it reaches the apparatus. The effect of heparin is cancelled by the addition of an antiheparin substance before the blood returns to the patient. This substance restores the coagulability of blood and therefore prevents bleeding. The maximum period for which an artificial kidney can be used is about four to six hours every two days. If kidney damage is temporary the patient is dialysed every second day until the kidneys function normally again. There are also patients who are being kept alive by means of an artificial kidney. In these cases permanent cannulas are inserted into a vein and into an artery and the patient must report for dialysing every second day or perform the dialysis at home.

CHAPTER 12

Heat and the Regulation of Body Temperature

Introduction

The temperature of any material is an indication of the molecular activity in the material (Chapter 8). If it feels "HOT", molecular activity is high. If it feels "COLD", molecular activity is low. There are, located in the skin, special receptor organs sensitive to changes in temperature, the "hot" and "cold" receptors. Changes in the surface temperature of the body over a wide area, as for instance when we feel hot or cold, or over a localised area as for instance when we touch something which is hot or cold, brings about a corresponding excitation or decrease in excitation, as the case may be, in molecular activity in these receptors. This change is translated by the receptors into nerve impulses which pass to the brain and are there interpreted as sensations of "hot" or "cold".

The rate of molecular activity in body tissues is most important, since the chemical reactions in cells only take place within narrow temperature limits. In humans the limits are a few degrees either side of 37°C, a temperature regarded as optimum for metabolic processes. The body temperature is maintained constantly at this optimum level by a very finely balanced, heat regulating mechanism. Fluctuations a few degrees either side of the optimum body temperature indicate a disturbance of this mechanism generally due to a disease process, and hence measurement of the body temperature is one of the first diagnostic tests performed on a patient.

Quantity of heat and temperature

Heat is the result of molecular movement caused by changing intermolecular forces (Chapter 8) and it is therefore a type of kinetic energy. The transfer of heat from a warm to a colder object is achieved due to the higher activity of the molecules of the hotter object. In order to understand heat production and temperature regulation, the two concepts of quantity of heat and temperature must be discussed. *Temperature* is a quantitative expression of the average rate of activity of each individual molecule in an object. It is independent of the size of the object. For example, a drawing pin can have the same temperature as a big block of iron. Temperature is measured in degrees Celsius (Chapter 2).

The total *quantity of heat* in an object, on the other hand, is the indication of the total activity of all the molecules of the object. The total quantity of heat of two objects differs, because it is dependent on the total number of molecules in the objects (i.e., their masses) and the average activity of each molecule of the object (i.e. their temperature). Let us assume that we have, as these two objects, a block of iron and a drawing pin, both at the same temperature initially. If we now immerse them in some water contained in a beaker, the block of iron will transfer much more heat to the water than the drawing pin, i.e., the temperature of the water will increase more in the case of the iron than in the case of the drawing pin. The unit that is used for the measurement of the quantity of heat is the joule (J). In order to gain an idea of a quantity of heat in one joule, it can be mentioned that about 4,18 kJ (4 180 J) is needed to increase the temperature of 1 kg water by 1°C.*

The manner in which different types of matter react to the addition of heat varies according to the substance concerned. The reason for this is that the structure of the different types of matter and therefore their intermolecular forces, differs. For this reason a small quantity of heat energy can cause a great change in the activity in certain materials. For example 4,18 kJ increases the temperature of 1 kg water by 1°C, and that of 1 kg copper by about 10°C. This difference is indicated quantitatively by the concept of *specific heat* of a material. Specific heat is the quantity of heat in joules that 1 kg of the material uses to change its temperature by 1°C. The units are joule/kilogram/°C (J/kg°C). According to this definition the specific heat of water is 4,18 kJ/kg°C and of copper about 1/10th of this energy, i.e., 0,389 kJ/kg°C. The higher the specific heat of a material the more heat energy it can absorb without excessively increasing the temperature of the material. Specific heat therefore plays an important role in the stabilisation of the temperature of the material concerned. This property is of the utmost importance in the body, where the temperature must always be kept at a constant

*Earlier the kilocalorie was used as the unit for the quantity of heat. In the SI this unit falls away. The most suitable unit to express energy values of foods and of energy requirements of a person is the kilojoule. The conversion factor from kilocalorie to kilojoule is 4,18. The energy values of a few foods in kJ are as follows:

	Kilocalorie/Gram	*Kilojoule/Gram*
Carbohydrates	4	16,8
Protein	4	16,8
Fat	9	37,8
Alcohol	7	29,4

The energy requirements of an adult performing light physical exercise are 3 000 kilocalorie i.e., 12 600 kJ.

level. It is therefore an advantage, from this viewpoint, that about two-thirds of the body consists of water, which has a high specific heat.

The quantity of heat energy gained or lost by an object can be determined by the following formula:

$$Q = CM\ (T_2 - T_1)$$

Q = quantity of heat in joules
M = mass of the object in kg
C = specific heat in J/kg°C
$(T_2 - T_1)$ = increase or decrease in temperature in °C

Example

A man, mass 70 kg, loses 84 kJ heat energy.

(*a*) If his normal temperature is 37°C, what will it be after the loss of heat?

(*b*) Compare this loss with the loss that occurs in an aluminium block of the same size, that has lost the same quantity of heat.

The specific heat of the body is the same as that of water, i.e. 4,18 kJ/kg°C, and the specific heat of aluminium is 0,915 kJ/kg°C.

(*a*)

$$\begin{aligned} Q &= CM\ (T_2 - T_1) \\ Q &= -84\,000 \text{ (loss)} \\ C \text{ water} &= 4\,180 \text{ J/kg°C} \\ T_1 &= 37°C \\ -84\,000 &= 4\,180 \times 70 \times (T_2 - 37) \end{aligned}$$

$$\begin{aligned} \therefore\ T_2 &= 37 - \frac{84\,000}{4\,180 \times 70} \\ &= 37 - 0{,}3 \\ &= 36{,}7°C. \end{aligned}$$

(*b*)

$$\begin{aligned} C \text{ aluminium} &= 915 \text{ J/kg°C} \\ -84\,000 &= 915 \times 70 \times (T_2 - 37) \\ T_2 &= 37 - \frac{84\,000}{915 \times 70} \\ &= 37 - 1{,}3 \\ &= 35{,}7\ C \end{aligned}$$

The temperature of the aluminium block has therefore fallen more than that of the man's body.

The production of heat

Because heat is the result of molecular activity, anything that increases molecular activity increases the production of heat. As a result of their continuous movement atoms or molecules collide with each other, resulting in a change of position of some of them. This causes a change in the intermolecular equilibrium. For example, an electrical current in a conductor changes the equilibrium and causes increased molecular activity. Similarly, chemical reactions cause changes in the forces and thus produce heat. The activities of the body are the result of hundreds of chemical reactions that take place continuously, resulting in a constant production of heat. In spite of this heat production the temperature of the body remains constant. This means that the body not only produces heat but also loses heat at the same rate. From the physiological point of view, the problem that the body has to solve is how to create a balance between heat production and heat loss. If heat production exceeds heat loss, the temperature rises, conversely, if heat loss exceeds heat production, the temperature will fall.

Heat production in the body. The chemical substances that serve as sources of energy are derived from the oxygen that we breathe and the nutrients that we eat. The reactions in the body responsible for heat production are also called metabolic processes. Part of the energy is used to build new tissues, to produce electricity in cells and to cause muscular contraction. The greater part of energy, however, is released as heat energy. Of the three primary nutrients, viz, proteins, fats and carbohydrates, fats produce the most energy, namely 38 kJ/g, while proteins and carbohydrates only produce 17 kJ/g. For this reason people living in cold climates consume more fats, and people eat more fat during winter than during summer.

Even when completely at rest a person produces energy in order to keep the basic activities of the body going. Such basic activities are the building of new tissues, the maintenance of tissues, the production of the resting membrane potential, the storage of reserve nutrient substances, the production and secretion of hormones, etc. The rate at which heat is produced as a result of these reactions under resting conditions is called the *Basal Metabolic Rate* (BMR). The following are a few factors that increase the rate of metabolism and consequently increase heat production.

1. *Hormones* that are secreted by the adrenal medulla and the thyroid gland.
2. *Muscular activity.*
3. *Shivering,* which is actually a subconscious and asynchronous muscular activity, especially designed to increase the amount of heat produced by muscles.

4. *Fever.*

Besides the above-mentioned factors that increase heat production in the body, the external environment may also play a part. Examples of this are heat from the sun, from warm objects such as heating appliances and from the various foods and beverages that have been consumed.

The total quantity of heat that is produced by the body during a certain period of time can be determined by means of a joulemeter. This apparatus is based on the principle that heat production and heat loss are in equilibrium in order to maintain the body temperature. The joulemeter consists of a special chamber with double walls, in between which water circulates. The temperature of the water can be measured at its input and output ends of the circulatory system. A person is placed in the chamber for a specific period. If the rate of water flow and the temperature difference between the two measuring points is known, the heat production per unit time can be calculated, since the heat produced is transferred through the walls of the chamber to the water.

When the person lies in a relaxed position on a bed, after having fasted for twelve hours, one can determine, by using the joulemeter his Basal Metabolic Rate (BMR). The normal value of the BMR for an average person with a mass of 70 kg is about 7 000 kJ per 24 hours. If the same person performs physical work the metabolic rate can rise as high as 12,500 kJ per day or even higher.

Heat loss

Heat can be lost by conduction, evaporation, radiation and convection.

(*a*) *Conduction.* By heat loss we mean the transfer of molecular kinetic energy from a place where the activity is high to one where it is low. The heat activated molecules collide with the inactive molecules and activate them. Each collision results in the active molecules losing part of their energy and transferring it to the less active molecules. This process is repeated until the molecular activity becomes the same throughout the whole of the object in question. The total quantity of heat remains the same, but it is now spread out. This means that heat flows from an area with a high temperature to an area with a lower temperature, until no temperature difference exists. The rate of heat conduction is different in various materials. In certain materials the molecular collisions cause only a minimal molecular movement and heat is conducted slowly. These materials are called heat insulators. The molecules of certain other materials on the other hand, are influenced easily by the molecular activity and such materials are called good heat conductors. The flow of heat is regulated according to the principles inherent in Ohm's Law as already described in Chapter 8.

(*b*) *Evaporation* means heat loss because the molecules that leave the liquid mass take their heat energy with them.

(*c*) *Radiation.* Objects also lose heat by the radiation of electromagnetic rays. This is the manner by which the sun sends its energy to the earth. When the rays reach an object they increase the molecular activity of the object, thereby increasing its temperature. Radiation is a result of molecular activity, i.e., temperature, and therefore all objects radiate heat to a greater or lesser extent. A hot object radiates more heat than it receives, and therefore actually loses heat by radiation. A cold object receives more heat than it emits and therefore gains heat. Radiation is also discussed in Chapter 13.

(*d*) *Convection* is the movement of a group of molecules from one place to another in order to transport heat. An example of this is the rising of air as it becomes heated. According to the gas law, discussed in Chapter 8, the intermolecular pressure in a gas increases with temperature. This means that this pressure exceeds the pressure caused by the weight of the air in the atmosphere and consequently the hot air moves upwards. In the case of movement caused by wind we would speak of forced convection. Forced convection causes very fast cooling of surfaces since it brings about a constant temperature difference between the surface and the air.

Regulation of body temperature

The production of heat and the loss of heat takes place continuously in the body. By balancing heat production and heat loss exactly, the temperature can be kept constant at 37°C. In FIG. 12.1 the processes of temperature regulation are summarised. Heat is produced by the metabolic processes in the body cells. This heat is transferred to the skin by (1) conduction through the tissues and (2) transport by the blood. Of the two mechanisms the latter is by far the most important, because the tissues, and especially fat tissue, are very poor conductors of heat. The skin loses heat to the surroundings by the processes of conduction, convection, radiation and the evaporation of sweat.

Keeping the temperature constant requires a very sensitive control mechanism that must be constantly informed about the production and loss of heat in every part of the body, since the activity of various parts of the body does not always proceed at the same rate. The control mechanism is situated in the brain and is called the *thermoregulatory centre.* The thermoregulatory centre receives information about the temperature of the peripheral organs, in the form of nerve impulses and through the medium of blood that flows through the centre. It integrates this information and adjusts heat production and heat loss accordingly, by, for example, changing blood flow

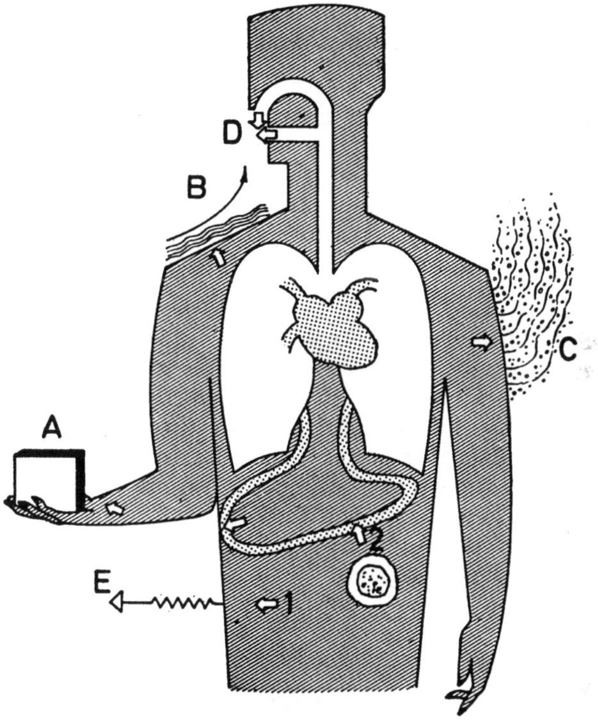

FIG. 12.1 Schematic representation of the temperature-regulating mechanisms of the body. 1 = Transport of heat in the circulatory system; 2 = Conduction of heat through the tissues; A = conduction of heat away from the skin; B = convection; C = evaporation of sweat; D = expiration of warm air; E = radiation.

through the skin or changing the rate of sweat secretion. In connection with sweat secretion, it must be noted that sweat only cools the body surfaces down if it evaporates, but not if it merely runs down the body.

The regulatory centre is so efficient that it maintains our body temperature within narrow limits. If the regulatory mechanism becomes damaged, or should it be incompletely developed, the temperature of the patient's surroundings will have to be kept constant artificially to ensure that his body temperature does not deviate from the normal. The thermoregulatory mechanism is especially weak in premature babies, and it is also less efficient in elderly people.

Diathermy

The effect of electrical shocks on muscles and nerves has already been discussed (Chapter 7). The frequency needed to stimulate muscle and nerve tissue is generally low. Higher frequency alternating currents can be used to heat up tissues. The electrical heating of a tissue is called diathermy, and can be applied surgically or therapeutically.

1. *Surgical diathermy*. When an alternating current with a frequency of 0,5-3,0 MHz is led through a knife-like electrode to a tissue, a small portion of the tissue at the point of the knife burns and disintegrates. The result is the same as if the tissue were cut through with a sharp knife, except that in diathermy the small blood vessels are burnt and blocked as they are cut through. Diathermy provides relatively bloodless surgery.

2. *Therapeutic diathermy*. If the electrodes do not touch the tissues but are only brought into close apposition to them, the tissues are heated but not burnt. In this case electromagnetic waves are produced (Chapter 13) which are sent through the tissues. The higher the frequency the greater is the heating effect.

Diathermy apparatuses are secondary output apparatuses. These apparatuses can be dangerous if the intensity of the radiation is too high, since the patient may be burnt. They must thus be handled with the utmost care.

SECTION IV

RAYS AND WAVES

CHAPTER 13

Electromagnetic Radiation

Introduction

The sun plays a special role in maintaining life on earth. Plants obtain energy from the sun through the process of photosynthesis, and they store it in the form of proteins, fats and carbohydrates. These substances in their turn serve as sources of energy for human beings and animals. The sun also provides light which enables man to make contact with the world around him through the medium of his eyes. It also provides warmth which is necessary for life.

The sun therefore provides *something* for the molecules of substances on earth which influences their activity. This *something* is known as electromagnetic radiation. Since it creates the impression of movement in straight lines it is called radiation. As radiation moves forward it exerts electrical and magnetic forces; hence the name *electromagnetic*. The two above-mentioned forces are actually not constant but vary in amplitude and direction. The graph (FIG. 13.1) depicting the variation of the forces is simplified to take the form of a sine wave, and hence this radiation is said to have wave properties.

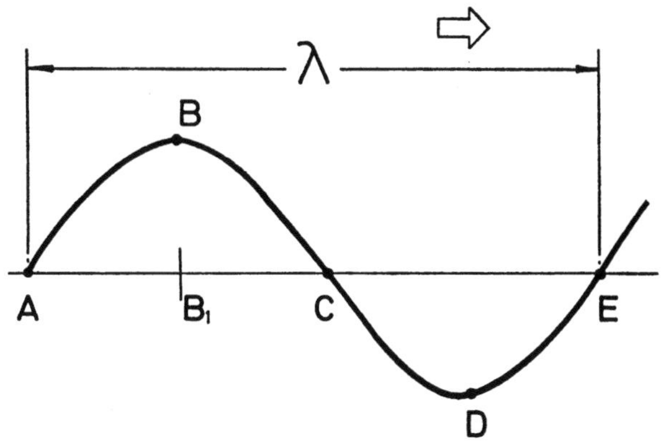

FIG. 13.1 Graphical representation of the variation of the electrical and magnetic forces exerted by the radiation as it moves forward. The amplitude of the forces is plotted against the distance moved by the radiation. λ = wavelength of oscillation.

Light waves and heat waves differ from each other in the frequency of their oscillations. There is actually a complete series or spectrum of electromagnetic radiation (waves) that is emitted by the sun (FIG. 13.2), and light and heat waves form part of this spectrum. Other electromagnetic radiations in the spectrum are X-rays, γ-rays, ultraviolet rays and radio waves.

Electromagnetic waves are produced whenever a charged particle is accelerated or decelerated. Electromagnetic waves travel through

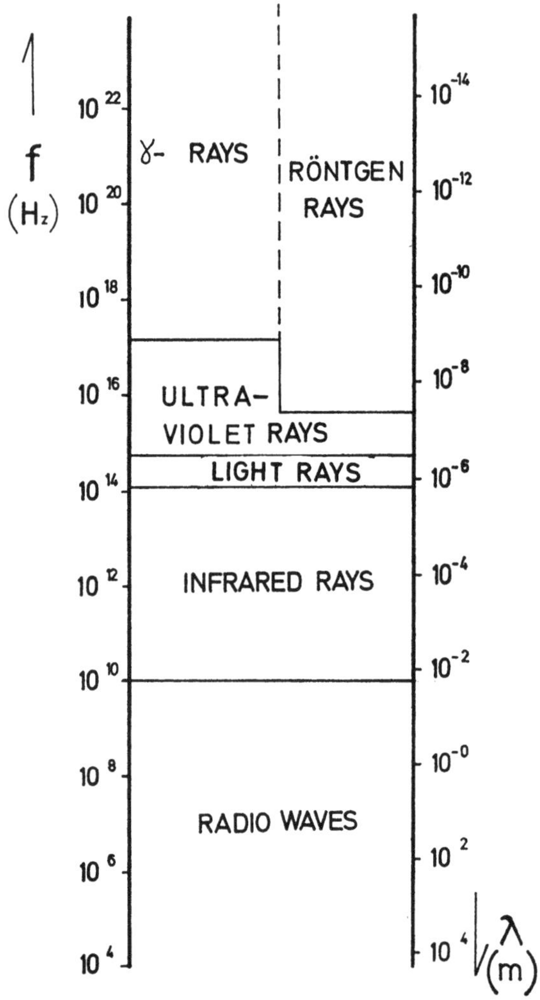

FIG. 13.2 The electromagnetic spectrum. f = frequency of radiation, λ = wavelength of radiation.

any space, and this is how they reach the earth from the sun. The speed at which the waves travel through air and a vacuum is 300 000 km/s. The waves are retarded, absorbed or reflected to varying degrees by different materials. The extent to which the radiations pass through a material or are reflected or absorbed by it, is determined by the frequency of the waves and the structure of the material.

We will only discuss here those sections of the electromagnetic spectrum that are of special importance in medicine. Sound waves do not form a part of the electromagnetic spectrum but are, however, discussed in this section, because many of the laws of electromagnetic waves are applicable to them.

General aspects

Frequency and wavelength. The number of times a wave oscillates in a second is called the *frequency* (f) of the wave. The unit of frequency is the *hertz* (Hz) (1 Hz = 1 vibration or oscillation per second). The distance covered by the radiation while completing one oscillation is called the *wavelength* (λ) (FIG. 13.1 — distance AE). Because the radiation is travelling forward at a constant speed, (300 000 km/s), the wavelength decreases with increasing frequency, i.e., the frequency (f) is inversely proportional to the wavelength (λ) or

$f = \frac{v}{\lambda}$; v = speed.

Example. If a wave has a frequency of 50 hertz, the wavelength is 300 000/50 = 6 000 km.

Inverse square law. The law that applies to the whole spectrum as well as to sound waves states that the intensity of the radiation, i.e., the number of rays per unit area is inversely proportional to the square of the distance from the source, if the source emits radiations in all directions. This inverse relationship is understandable if we consider that the radiations spread out away from each other as they move away from the source (FIG. 13.3).

The area (A) on which the radiation falls is the surface area of a sphere around the source, and this increases with the square of the radius (R) i.e., $A = kR^2$. The number of radiations per unit area (I) is therefore the total number of radiations (L) sent out in all directions divided by the area ($A = kR^2$), i.e.,

$$I = \frac{L}{kR^2}$$

The importance of the intensity of radiation can be shown by the following simple example. In order to read what is written here, light is required. The amount of light falling on the area to be read is important, but not the total amount of light in the room where we happen to be.

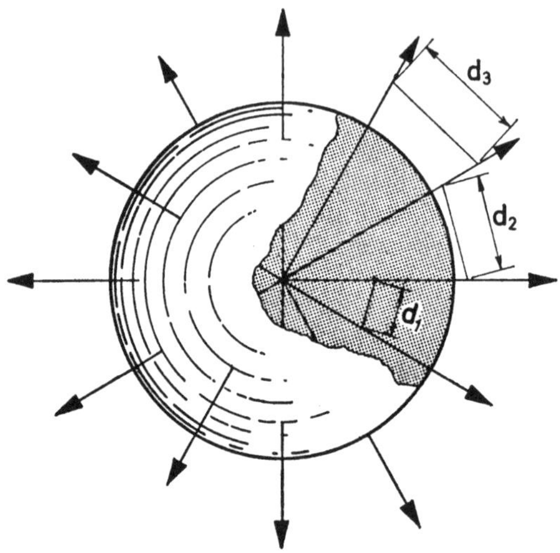

FIG. 13.3 Radiation spreads out in all directions. The distance between the rays increases the further they move from the centre.

Heat waves

Heat radiation has already been mentioned in connection with the temperature-regulating mechanisms of the body (Chapter 12). Heat is increased oscillatory activity of the molecules concerned. During oscillation, the charged particles in the molecules are alternately accelerated and decelerated, and as a result electromagnetic waves are produced. These waves move through air and vacuum, are reflected by light-coloured surfaces and are absorbed by dark-coloured rough surfaces. Absorption of radiation by objects means that the energy of the radiation is transferred to the molecules of the object. The molecular movement and therefore the heat content of the object, increases. In the process heat is transferred from one object to another.

In the manufacture of our clothing we make use of the ability of certain materials to reflect radiation and of others to absorb radiation. During the summer we wear light coloured clothes since they reflect the heat of the sun better than darker clothing. For the same reason, dark materials are used in winter because they absorb heat better.

All objects radiate heat and the hotter the object, the more heat is radiated. The human body produces heat, part of which is radiated as heat rays, but it also receives heat waves from its surroundings. When the temperature of our body is higher than the temperature of the surroundings, it will radiate more heat than it receives and thus it will lose heat. Similarly, the body gains heat from the surroundings if its temperature is lower than that of the surroundings.

Heat radiations are not uniform as far as their frequencies are concerned, but consist of a series of frequencies in the infrared region of the spectrum. The higher the temperature of the object, the higher the frequency of the radiation emitted by it. Infrared radiation is produced for therapeutic purposes by special lamps and is used to heat tissues and thus to increase their activity.

Light waves

Light waves are those waves of the spectrum which have a special effect on the eye. When light reflected from objects falls on the retina, it is absorbed and initiates certain chemical reactions. This stimulates the nerves of the eye, which send impulses to certain areas of the brain. In this way the brain of the person sees the image of the object.

Light is produced by the sun, but it can also be produced artificially by the movement of electrons in special metals or gases. Two well-known methods used for the production of artificial light are the filament lamp and the neon lamp. In the filament lamp, electricity is passed through a special metal such as tungsten. The movement of the electrons in the metal disturbs the intermolecular forces, and as a result of this, the electrons alternatively accelerate and decelerate, and light waves are emitted. The lamps simultaneously produce large quantities of heat which in this case serve no useful purpose. This means that the amount of light produced is relatively low for the amount of electricity that is supplied; thus the efficiency of the lamp is low. In neon lamps the electricity is passed through a gas contained in a tube of the lamp. The electrons are removed from the molecules of the gas and become accelerated. Neon lamps are more efficient than filament lamps, since they produce more light and few other radiations per unit of electricity.

Different materials reflect or absorb light falling on them to different degrees. We see objects when the light reflected by them reaches our eyes. The amount of light that reaches our eyes from the different points of an object is determined by the colour and texture of the material of the object. Just as in the case of heat waves, light-coloured surfaces reflect more light than dark-coloured surfaces, while dark-coloured surfaces absorb more light. This means that the eyes receive more light from light-coloured objects than from dark-coloured objects. It is this property that makes it possible to distinguish between dark and light colours or materials and to recognise objects.

Like heat waves, light waves constitute a part of the electromagnetic spectrum. The different frequencies of light may be recognised as different *colours*. The white light of the sun is composed of a mixture of light waves of different frequencies. The different frequencies can be separated from each other by the use of prisms (FIG. 13.4). Prisms have the ability to bend light rays, and the extent to which the rays are bent is dependent on the frequency of the radiations. In this way radiations with different frequencies can be separated from one another. A well-known example of this is to be found in the rainbow. In this case the droplets of moisture in the atmosphere serve as prisms separating the light rays which we are then able to recognise as the different frequencies of the visible spectrum. The colours vary from red to violet, which represent the lowest and highest frequencies of the *visible spectrum*, respectively.

More laws dealing with light radiation will be discussed in Chapter 14.

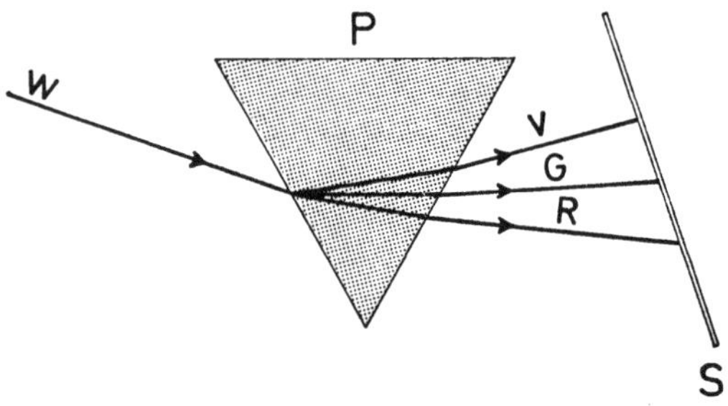

FIG. 13.4 Schematic representation of the principles according to which prisms separate rays having different frequencies. White light (W) consists of waves of various frequencies. The prism (P) refract violet (V) rays more than green (G) and red (R) rays. On the screen (S) the colours are thus separated.

Ultraviolet radiation

The rays with higher frequencies than the violet rays of the visible spectrum are called ultraviolet rays. Most of the ultraviolet rays from the sun are absorbed by the atmosphere, and only a small proportion of these reaches the earth. Ultraviolet radiations have a considerable effect on the human body. They are absorbed by the skin and can therefore only penetrate the body to a small extent. In the skin the ultraviolet radiation stimulates the pigment cells to produce a *pigment*, known as melanin which protects the body from further ultraviolet radiation. After exposure to the sun, this pigment imparts to the skin a brown colour. This protective mechanism is important since excessive exposure to ultraviolet radiations destroys living cells. If the intensity of the ultraviolet radiation is high or if the body is exposed to the radiation for a long time, the cells of the skin are destroyed. The destructive effect of ultraviolet radiation can be turned to useful purpose to destroy bacteria in operating theatres, and to sterilise instruments.

Ultraviolet radiation of lower intensities play a role in the *production of active vitamin D.* An inactive chemical substance is present in the skin, and this changes into active vitamin D under the influence of ultraviolet rays. Vitamin D is absolutely essential for normal calcium metabolism and for the calcification of the bones. The bones of children who do not receive enough sunlight do not calcify properly and as a result become weak. The condition that develops is called rickets. It can be remedied by giving additional vitamin D, or by exposure to sunlight, or artificial ultraviolet radiation. Ultraviolet rays can be produced artificially by neon or mercury lamps.

X-rays

X-rays are produced when electrons moving through a vacuum are suddenly halted by a metal plate. The collision decelerates the electrons abruptly and electromagnetic waves of a certain frequency (X-rays) are emitted. X-rays, as in the case of light rays, have an effect on specially-prepared photographic plates. Depending on the amount of radiation falling on the plate, the plate may show different degrees of density after development. When the body is placed between the source of the X-rays and the photographic plate, the rays penetrate some of the tissues and are absorbed to a greater or lesser degree by other tissues. Bones absorb the radiation to a high degree, and if a part of the skeleton is placed between the source and the plate, the part of the plate behind the bones receives less radiation and appears in outline as a less dense shadow. This principle is used to investigate bone and other dense structures in the body and to detect abnormalities in structure. X-rays play an important role in the diagnosis of a variety of illnesses and are also useful for the detection of foreign objects in the body.

Tissues and structures which normally allow X-rays to pass through can be studied by making use of special substances, which absorb the radiations. Such substances are called contrast media. The contrast medium is introduced into the structure or organ under observation, and since it absorbs the X-radiation, a photographic image of the organ or structure is obtained. Bariumsulphate solution is used for example, to investigate the contours of the alimentary canal. Similarly, organic iodine compounds are used for the radiological examination of the liver, gallbladder, kidney and air passages. The photographic plate is a permanent record of the examination.

With the aid of a special fluorescent screen, organs can be seen while they are being irradiated. The screen produces light rays when X-rays reach it from behind. By this method movements of the organs, for example heart contractions or bowel movements, can be investigated. The method is called *fluoroscopy*.
X-rays have a destructive effect on living organisms and may therefore be used to destroy abnormal cells such as tumour cells in the body. Continuous or repeated exposure to X-rays also destroys healthy cells, however. Persons, such as doctors and nurses, who are frequently exposed to X-rays, must wear lead screens to avoid dangerous excessive exposure. A few millimetres of lead absorb X-rays completely, and therefore protect the body from the results of chronic exposure to X-rays.

Radio waves

Radio waves have lower frequencies than infrared waves, and are produced in metal transmitter antennae by a fast to-and-fro movement of electrons. The radiation can be sent out in all directions or can be concentrated in a specific direction. It moves through air and vacuum and can again be received by metal antennae. When the waves reach the receiving antenna, they cause the same electron movement that produced them in the transmitter. Radio waves are thus used to transmit information between the two antennae that may be situated at very long distances from one another. Radio waves are used in medicine in so-called *telemetering instruments*. These instruments are, for example, important in space travel. They can be used to transmit radio waves, for instance, every time the heart of the astronaut beats. A receiver on the earth registers these waves, and the heart rate of the space traveller can therefore be calculated. Similarly, the amplitude or the frequency of the radiations can be varied by the blood pressure of the astronaut. The receiver on earth picks up the waves and interprets them as a recording of the blood pressure of the astronaut. If the amplitude of the waves is varied according to some information to be transmitted, we talk of amplitude modulation (AM), and if it is the frequency of the radiations which is modulated, we talk of frequency modulation (FM).

Gamma radiation

Gamma radiation is emitted when certain radioactive elements disintegrate. This type of radiation is discussed in detail in Chapter 15.

Lasers

Normal light produced by lamps spreads out in all directions and is composed of a whole series of waves, the frequency of which differs in each case. An artificial ruby crystal can produce a single frequency (deep red) of radiation emitted in parallel rays if normal light is passed through it. These rays are called laser rays (LASER = *L*ight *a*mplification by *s*timulated *e*mission of *r*adiation).

Because these rays are concentrated, they can be used to destroy small amounts of tissues. Lasers are specially well absorbed by pigment cells, such as the pigment layer of the retina in the eye. With the aid of laser rays the surgeon can operate on the retina. Irradiation lasts for a few milliseconds and no narcosis is necessary for the operation. The same procedure can be used to destroy malignant tumours.

CHAPTER 14

Light and Sight

Introduction

Light, or the visible spectrum is the name given to that portion of the electromagnetic spectrum which stimulates the eye and thus produces *light sensations* in the brain. The influence of light waves with different frequencies on the brain cells is interpreted as different colours. A mixture of rays containing all the frequencies of the visible spectrum is interpreted by the brain as *white.* An object is perceived when the rays reflected by it reach the eye. Light rays can be reflected, refracted and absorbed. The laws governing these processes are collectively called the Laws of Light.

Laws of light

As it has already been mentioned (Chapter 13), light rays may be considered as rays moving in a straight line. When the rays reach an object, they can either pass through it, or they can be absorbed or refracted by it. Usually these three processes all take place simultaneously, i.e., some rays are reflected, some are absorbed and some are refracted. When a person looks at an object, the amount of light reaching the retina of his eyes is determined by the relative contributions of each of the above-mentioned processes. A surface which reflects a large percentage of the light will be seen as a light-coloured surface, while a surface which absorbs most of the light will be seen as a dark surface.

(*a*) *Reflection and Mirrors*

Mirrors are constructed in such a way that they reflect most of the rays reaching them. They are useful therefore in explaining the laws governing reflection. A mirror consists of a glass plate, the back surface of which is painted a dark colour (FIG. 14.1). Rays from an object (ABC) reaching the mirror will be reflected in such a way that the incident and reflected rays produce the same angle (α) at the mirror surface. As a result of the reflected rays reaching the eye of the observer, an image of the object (A'B'C') is seen in the mirror.
The ability of a mirror to reflect rays means that the direction of the incoming rays can be altered in such a way that objects which are

outside the field of vision are also seen. In medicine eye specialists, ear, nose and throat specialists, urologists, dentists, gastrologists, etc. make use of this principle to examine small inner structures of the mouth, stomach, throat, nose, etc., which normally lie outside the field of vision.

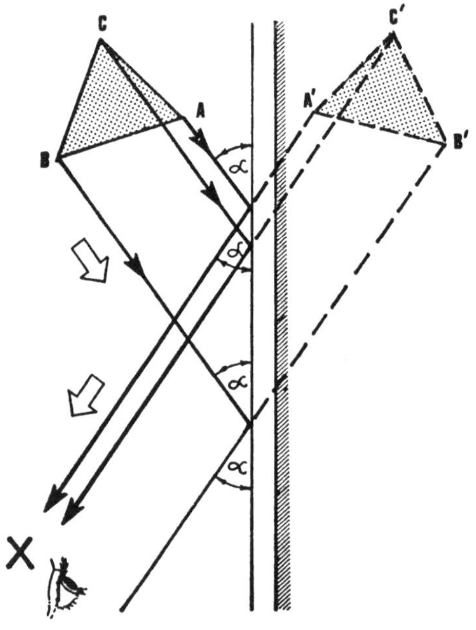

FIG. 14.1 Schematic representation of reflection in mirrors. ABC = object, A'B'C' = image of the object, X = eye of the observer.

Convex and concave mirrors. When a curved glass plate is used for the construction of a mirror, a convex or a concave mirror is obtained, depending on whether the inner or the outer surfaces are the reflecting surfaces. In a convex mirror the reflecting surface is on the outside of the curve (FIG. 14.2a), and in a concave mirror it is on the inside (FIG. 14.2b). When parallel rays reach a convex mirror, the direction of the rays is changed in such a way that they move away from each other, i.e., they *diverge* as if they were all emerging from a single point. On the other hand, when parallel rays reach a concave mirror, the direction of the rays is changed in such a way that they all pass through a single point, i.e., they *converge*. This means that a concave mirror concentrates the rays in one point. A piece of paper held at this point may catch alight if the intensity of the rays is strong enough. For this reason this point is known as the *focal point* or *focus* (FOCUS = hearth, Latin).

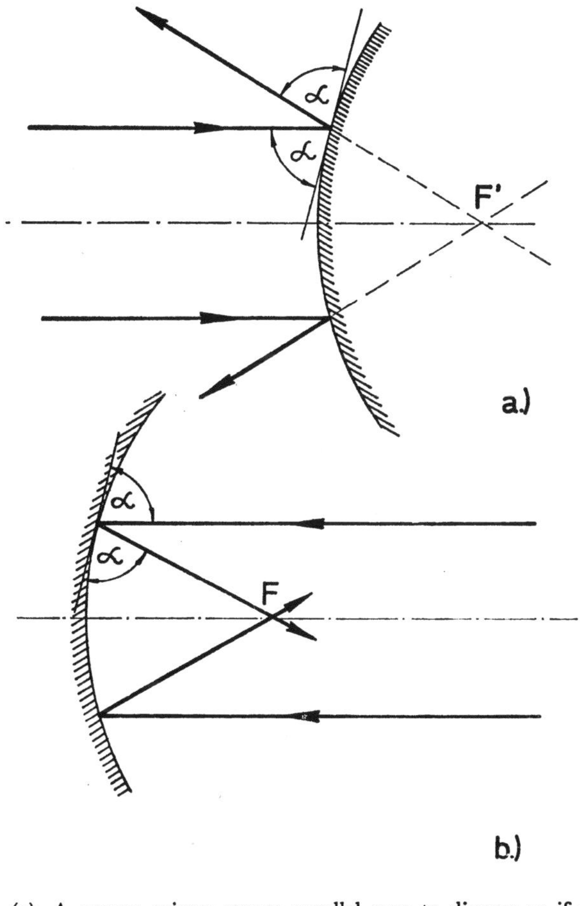

FIG. 14.2 (*a*) A convex mirror causes parallel rays to diverge as if they were coming from an imaginary point (imaginary focal point, F[1]).
(*b*) A concave mirror concentrates parallel rays in a single point (focal point, F).

Endoscopes. An endoscope is an instrument which is used in medicine to investigate the inner structures of the body. Endoscopes used for different purposes work on the same principles of reflection. The auroscope is used to examine the ear cavities, the head mirror the throat and nose cavities, the ophthalmoscope the structure of the eye, the cystoscope the bladder, the gastroscope the stomach and the proctoscope the rectum and large intestine. In these instruments (FIG. 14.3) the light sent out by the source is concentrated with the aid of a concave mirror on the structure to be examined. The structure reflects the rays which then reach the eye of the observer through an opening in the mirror.

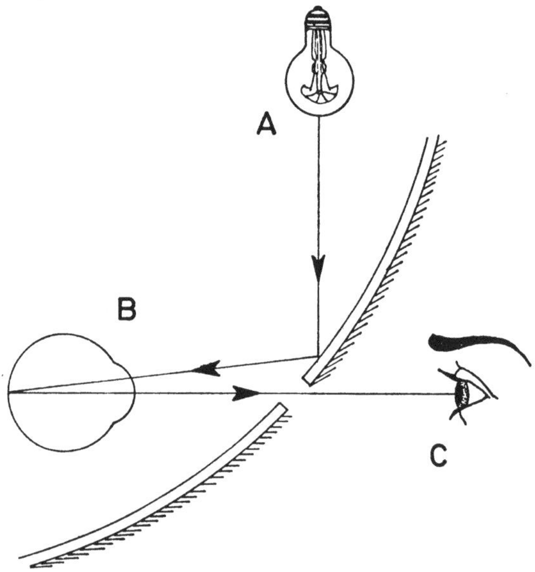

FIG. 14.3 The principles of endoscopes. A = Source of light; B = the structure to be examined (in this case the retina of the eye); C = eye of the observer.

One-way Mirrors. One-way mirrors reflect light in the normal way from their front surface, and in addition a person behind the mirror can see objects which are in front of the mirror. These mirrors are used for example in hospitals to observe patients who are seriously ill without them being aware of it. The dark paint on the back surface of the mirror which normally absorbs the light has openings which allow a small amount of light to pass through (FIG. 14.4). A person C standing in front of the mirror perceives many reflected rays from objects (e.g. A) which are in front of the mirror. The small amount of light passing through from the back of the mirror is masked by the reflected light and therefore the person in front cannot see the objects behind the mirror. A person D standing behind the mirror receives light only from the front since the dark back surface of the mirror absorbs the rays reaching it from behind.

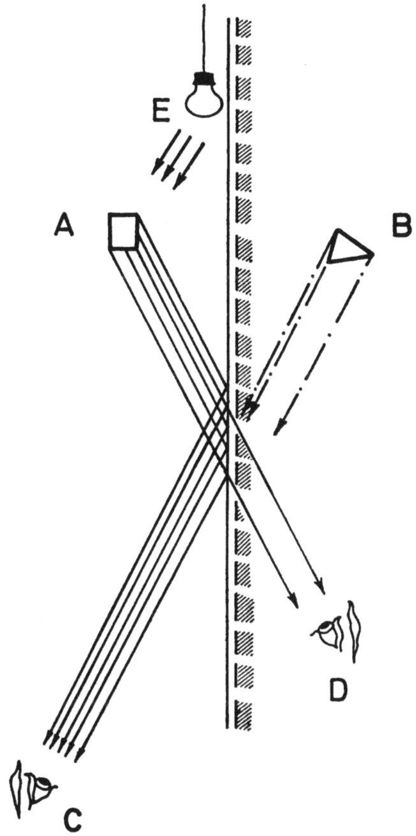

Fig. 14.4 Principles of one-way mirrors (see text).

One-way mirrors work optimally when the area in front of the mirror is well illuminated and the area behind the mirror has poor illumination. The principles of the one-way mirror are also applied in the construction of windows for modern buildings. In this case the reflecting surface is on the outside making it difficult to see through the window from the outside. The people inside the building on the other hand, still receive light and can see what is happening outside. These types of windows are also used in supermarkets for surveillance of the customers, in order to detect shoplifters.

(*b*) *Refraction*

Refraction results from the fact that different materials do not conduct light rays at the same rates. Water and glass, for example, conduct light rays at a slower rate than air. Light rays become refracted when they pass through one material into another, provided that the rays reach the dividing surface between the two materials at an angle other than a right angle (FIG. 14.5). Let us assume that two rays, A and B, leave the source at the same instant and move parallel to each other, first in air and then in water. After a certain time the two rays are at positions C and D. The ray AC enters the water first and therefore moves slower. It means that this ray will move through a shorter distance (CE) than the other ray (DF) still moving through air. After the ray BD has also reached the

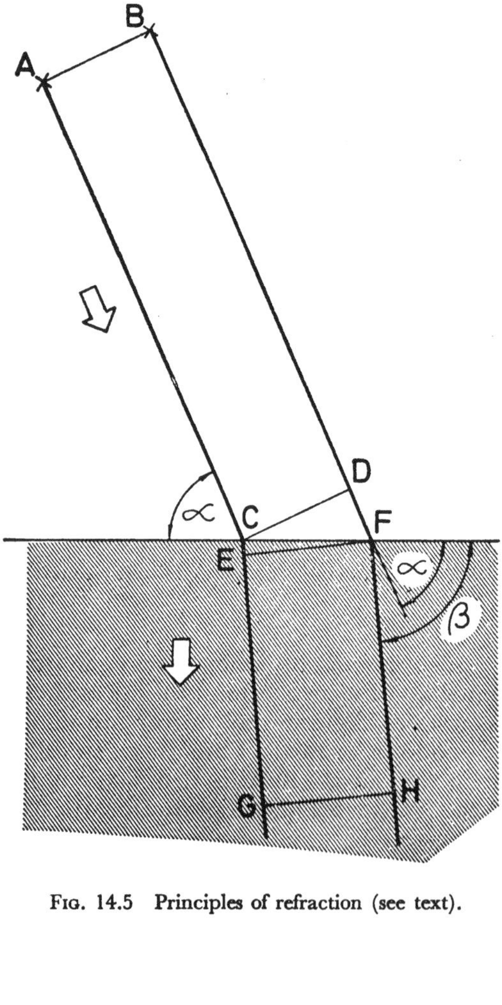

FIG. 14.5 Principles of refraction (see text).

water level, both of the rays will move at the same speed in the water and reach the positions G and H at the same time. The direction of the rays has therefore changed from direction AC to direction CG. Assume that the angle between the incident ray AC and the dividing surface is α, and the angle between the refracted ray and the dividing surface is β. Generally β is greater than α if the speed of light in the second material is less than in the first, as is the case in the above example (Fig. 14.5). The two angles are the same if the speed in the two materials is the same, and β is smaller than α if the speed in the second material is greater than in the first. This means that the degree and direction of refraction is determined by the speed of light passing through the materials.

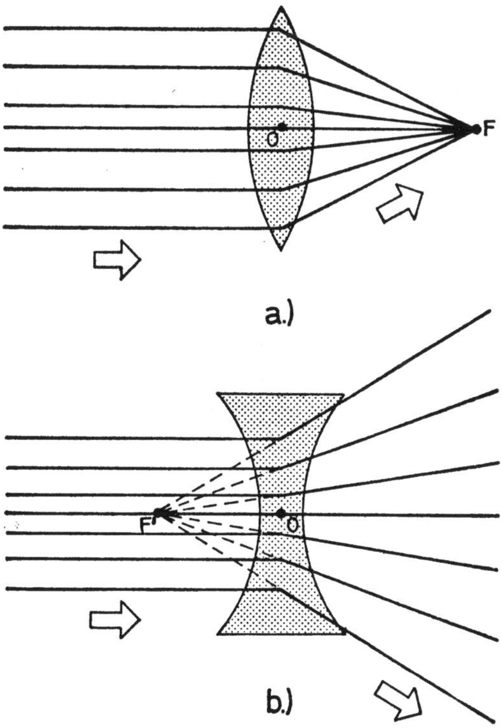

Fig. 14.6 (*a*) A convex lens concentrates parallel rays in a focal point, F.

(*b*) A concave lens disperses parallel rays as if they were coming from an imaginary focal point, F^1).

Lenses

In lenses the principles of refraction are applied to disperse or to concentrate rays. A whole series of lenses is available, but only *biconcave* and *biconvex* lenses will be briefly discussed here. In the case of biconvex lenses, both surfaces are *rounded* (FIG. 14.6b) and in

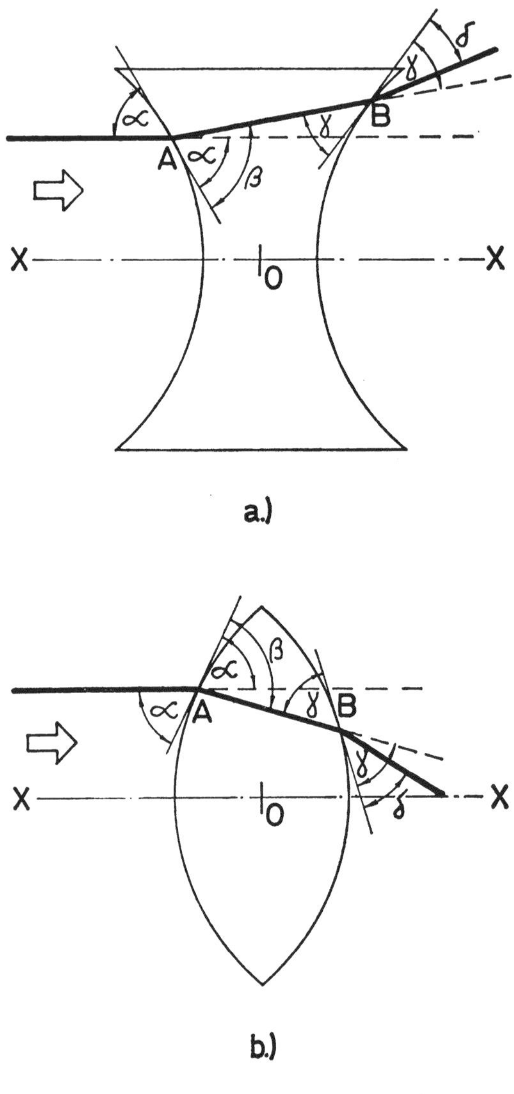

FIG. 14.7 Principles of refraction by biconcave (a) and biconvex (b) lenses. (See text.)

the case of biconcave lenses both surfaces are *hollow* (FIG. 14.6a).
Refraction through lenses. The manner in which a light ray moves through the above-mentioned lenses is shown in FIG. 14.7. It must be remembered that the speed of the light ray decreases when it leaves the air and enters glass or liquid, and therefore the angle which the ray makes with the dividing surface increases. On the other hand, when the rays leave the glass and enter air, the angle becomes smaller. For the sake of clarity the terms *axis of the lens* and the *centre of the lens* are defined. For the purposes of this discussion it will be assumed that the two curvatures of a lens are the same. The axis of the lens is the straight line joining the central points of the curves (line XX in FIG. 14.7). The centre of the lens (point O in FIG. 14.7) is the point bisecting the section of the axis between the two curvatures. Concave and convex lenses will be discussed together.

Let us assume that a ray moving parallel to the axis of the lenses reaches the surface of either lens at point A and makes an angle α with the surface of the lens concerned at this point.* Since the ray has penetrated the glass from an air medium, the angle β at which it leaves the surface will be greater than the initial angle. The ray now reaches the second surface at point B and makes an angle γ with this surface. Since the ray is leaving a glass medium to pass into air, the angle it makes on leaving (δ) is smaller than the angle γ. In the biconcave lens the rays are refracted away from the axis and in the biconvex lens they are refracted towards a point on the axis. If the light rays are strong enough they are able to ignite a piece of paper placed at this latter point. The point is called the focal point (F) of the lens, and the distance from the centre (O) of the lens to the focal point is the focal length. In biconcave lenses the light rays disperse as if they were coming from the same point (F′) on the axis. This point is called the imaginary focal point.

Refracting power of lenses. The extent to which rays are refracted by a lens is dependent on the material from which the lens is made and the curvature of the lens. As has already been mentioned, the extent to which the rays are refracted is dependent upon the difference in rate of travel of the rays through the materials concerned. The greater the difference in rate, the more the rays are refracted. The curvature also influences the refraction: the *fatter* the lens, the greater the refraction of the rays, i.e. the shorter the focal length (FIG. 14.8). The extent of the refracting power of lenses is measured in *dioptres.* The smaller the focal length, the greater the refracting power, i.e.,

$$\text{Refracting power} = \frac{1}{\text{focal length (m)}}$$

*In two-dimensional drawings lenses are represented by curves. The orientation of the surface at a certain point is represented by the tangent to the curve at that point.

This means that if the focal length is 1 metre, the refracting power is 1 dioptre; if the focal length is half a metre, the refracting power is 2 dioptres. The refracting power is inversely proportional to the focal length.

The principle of refraction of rays in lenses is applied in the magnifying glass (FIG. 14.9). A magnifying glass is a biconvex lens, with a short focal length, i.e., the lens has a great refracting power.

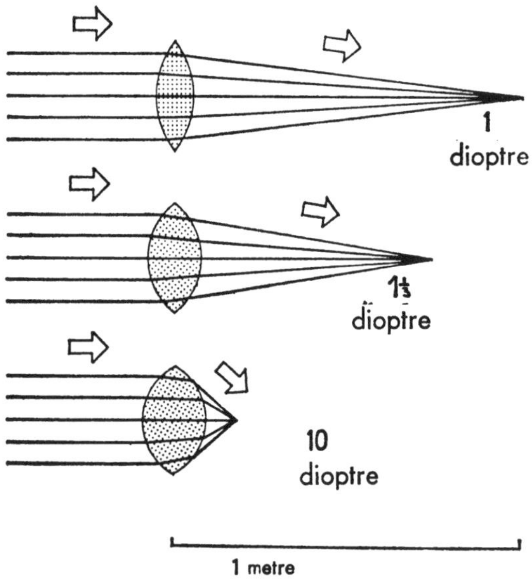

FIG. 14.8 Schematic representation of the connection between the curvature and focal length of lenses.

For example, let us assume that an object (AB) is placed between the lens and the first focal point (F_1). Parallel rays AA′ and BB′ reaching the lens from the object are refracted in the direction of the second focal point F_2, and the observer situated at this point sees the rays as if they were coming from the enlarged image A″ B″.

Image formation in lenses. Biconvex lenses possess the ability to form two-dimensional images of objects. This is possible since all the light rays which come from a point in front of the lens and move through it, converge again behind the lens *provided that this point is not closer*

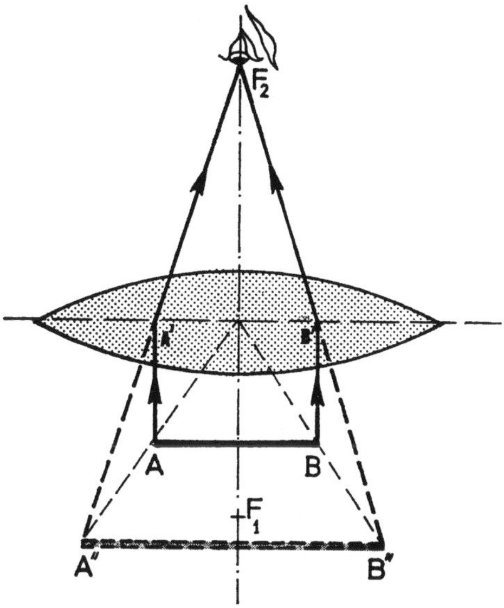

FIG. 14.9 The principles on which the magnifying glass works (see text).

to the lens than the focal point (FIG. 14.10). This means that every point of an object which is placed in front of the lens has a corresponding image behind the lens. If a screen is placed at this point, the image of the object can be seen upon it. The size of the image may be smaller, may be the same size or may be larger than the object, depending upon the distance between the object and the lens. The position and size of the images can be determined by an easy graphical method if the focal length (f) of the lens is known.

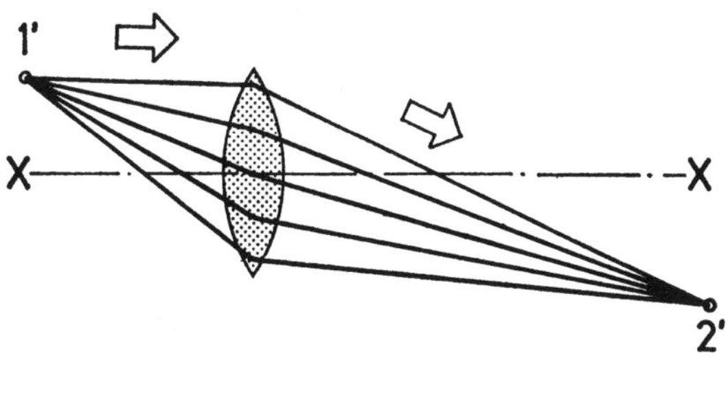

FIG. 14.10 Lenses have the ability to concentrate all rays coming from one point (1′) at a second point (2′).

The following are important in understanding the graphical analysis (FIG. 14.11a).

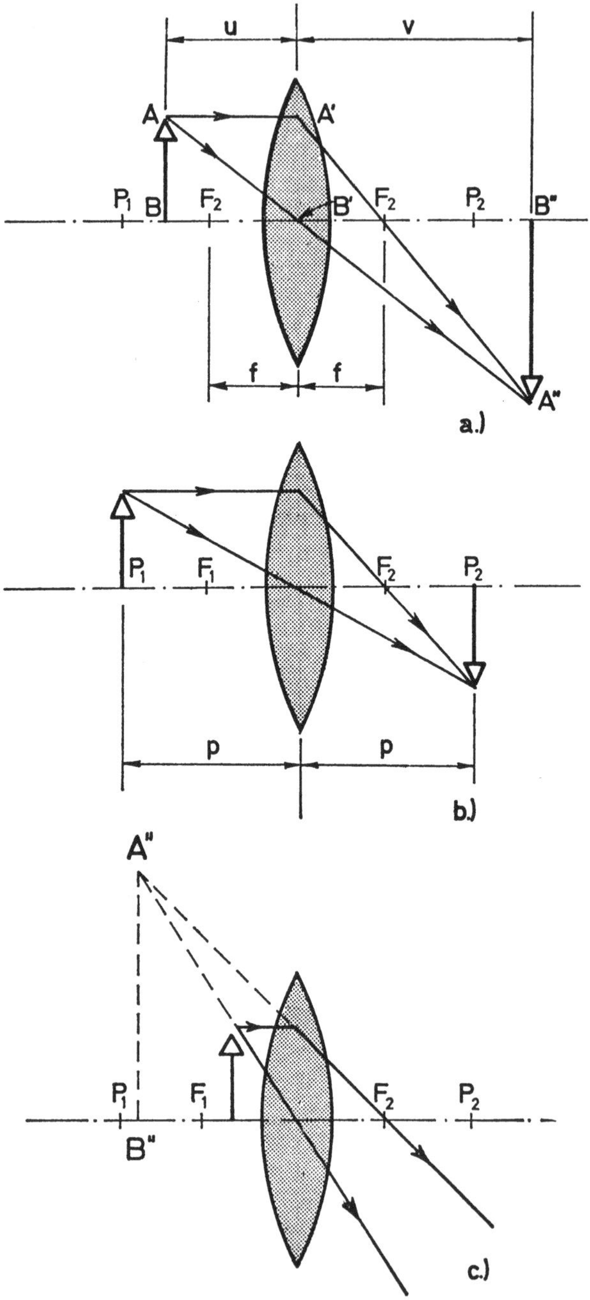

FIG. 14.11 The relation between the object and the image formed by a biconvex lens (see text).

1. Parallel rays are refracted through the focal point. The direction of ray AA′ is changed to $A'F_2$.
2. The rays which pass through the centre of the lens are not refracted at all — ray AB′.
3. The point of intersection of any two rays coming from the same point determines the position of the image, because all rays from the same point will eventually pass through one point.

Using the above-mentioned points we can demonstrate that when the object is at a distance twice the focal length from the lens, the image is the same size as the object and at the same distance from the lens (FIG. 14.11b). For this reason the names *isometric point* and *isometric length* respectively, are given to the above-mentioned position and distance from the lens of the object and the image (ISO = same). The relationship between the focal length (f) and the isometric length (p) is thus p = 2f. If the object (AB) is nearer to the lens than the isometric point (P_1), but is further away than the focal point (F_2), the image A″B″) will be enlarged (FIG. 14.11a). If the object is further from the lens than the first isometric point, the image will be smaller and will lie between the focal point and the second isometric point (prove this with the aid of a diagram). Notice that in all of the above-mentioned cases the image is inverted. As mentioned earlier, there will be no image if the object is nearer to the lens than the focal point (FIG. 14.11c).

Optical apparatus

In practice the principles of image formation are applied in cameras, film projectors and microscopes.

(*a*) *The camera*
In the camera the object is further from the lens than the isometric length and consequently a small image is formed on the photographic paper placed between the focal point and the isometric point on the other side of the lens. The image on the paper is developed and fixed by chemical methods.

(*b*) *Projection apparatus*
In a projection apparatus the object, in this case the film, is placed between the focal point and the isometric point. An enlarged image is formed on the screen which is further from the lens than the isometric point.

(*c*) *The microscope*
The microscope is an optical instrument constructed of two lenses each having great magnifying power (FIG. 14.12). The object (AB) is placed between the focal point (F_X) and the isometric point (P)

of the bottom lens (XX), the objective. As a result of this arrangement an enlarged and inverted image (B′A′) of the object is formed behind the lens. The image so formed serves as an object for the second lens, the ocular (YY), which works as a magnifying glass.

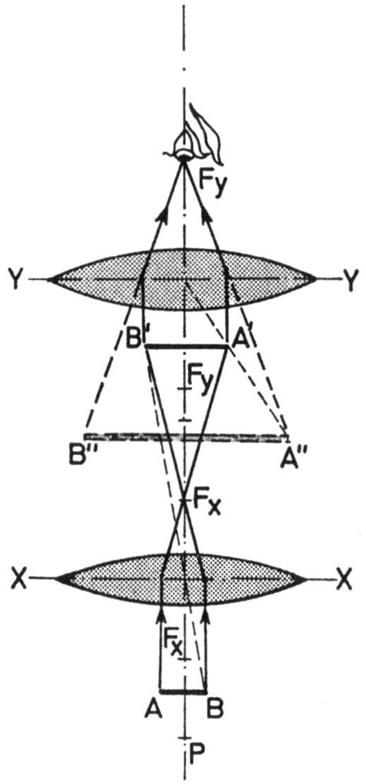

FIG. 14.12 Schematic representation of the principles on which the microscope is based.

This means that the object (B′A′) is closer to the ocular lens than its focal point. The observer sees the rays as if they were coming from the image (B″A″). The magnification obtained with the microscope is the product of the magnification obtained with each of the lenses.

Microscopes are used extensively in biology and medicine to examine tissues and cells. Magnification of up to 3 000 times is possible. This means that with this great magnification an object 10 micrometres in diameter shows up as an object 30 millimetres in size. Magnifications of 100 000 times or more can be obtained with the electron microscope. Such magnifications make the examination of small living organisms such as viruses and parts of cells possible.

The eye

From a practical point of view the eye may be regarded as consisting of a lens which refracts incoming rays in such a way that an image of the observed object is obtained on the retina (FIG. 14.13). The light falling on the retina initiates chemical reactions which stimulate the nerves of the eye. The impulses produced are transmitted via the nerves to the sight centres in the brain where an image of the object is reproduced.

In the case of an ordinary glass lens the distance of the image from the lens is dependent upon the distance of the object from the lens. In the eye the retina is at a fixed distance from the lens, and the

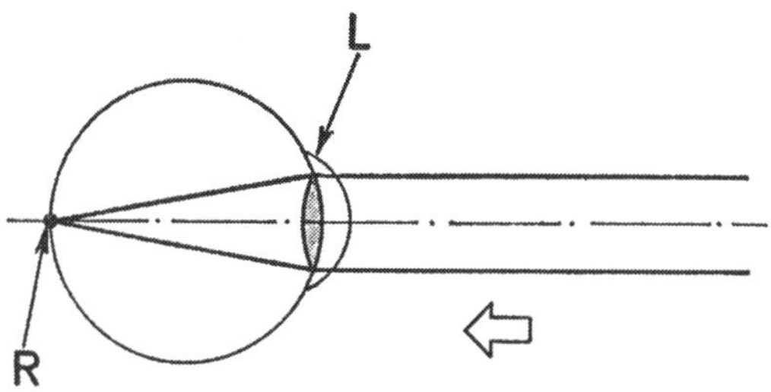

FIG. 14.13 The lens (L) of the eye refracts parallel rays in such a way that they become concentrated on the retina (R).

image must always fall on the retina. To make this possible the refracting power of the lens changes according to the distance of the object from the lens. As mentioned earlier, the ray-refracting ability of the lens is dependent upon the curvature of the lens (FIG. 14.8) and the eye makes use of this principle in order to form the image on the retina, whatever the distance of the object from the lens.

A lens concentrates parallel rays at the focal point (14.14a). In the normal relaxed eye the curvature of the lens is such that the focal point is on the retina. When one looks at a closely situated object (nearer than 6 metres for a normal eye), the image will be formed between the focal point and the isometric point, i.e., it will fall behind the retina, and the object will not be seen clearly without an adjustment in the curvature of the lens (FIG. 14.14b). If the curvature and thus the refracting power of the lens increases, both the focal point

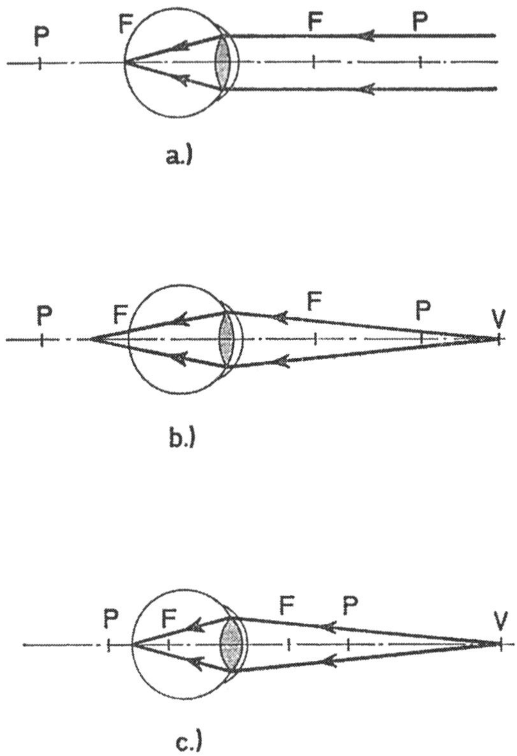

FIG. 14.14 The mechanism of accommodation in the eye (see text).

and the isometric point will be closer to the lens, and with the right amount of refraction the image will fall on the retina (FIG. 14.14c). The ability of the eye to alter the refracting power of the lens is called the *accommodating ability* of the eye.

The lens of the eye consists of transparent elastic fibres enclosed in a strong elastic membrane (FIG. 14.15). The elastic membrane is connected to membranous ligaments at the sides of the lens. The ligaments are connected to a ring-shaped muscle situated around the outer surface of the lens. In the normal eye this muscle is in a relaxed state. The tension in the ligaments holds the lens in a flattened position. As a result of this objects further away from the eye than 6 metres can be seen clearly by the eye at rest. Under these circumstances the rays from the object are parallel rays, and the image is formed at the focal point, i.e., on the retina. When one looks at objects which are closer to the eye than 6 metres the incoming rays

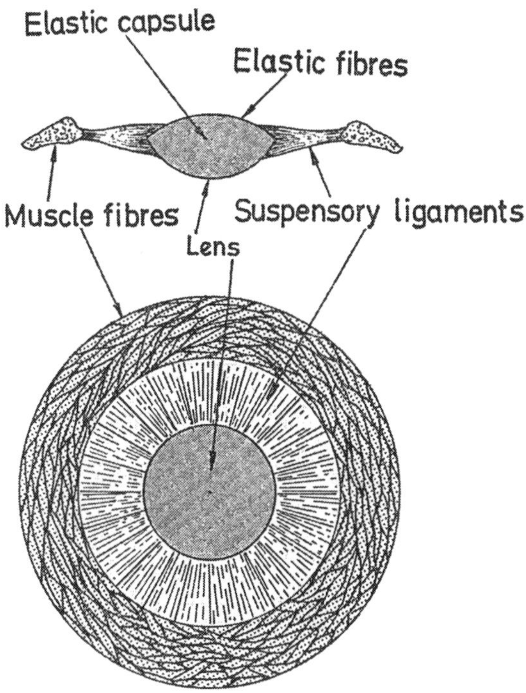

FIG. 14.15 Schematic representation of the accommodation apparatus of the eye.

must be refracted more and more, the closer the object comes. To produce greater refraction of the rays the curvature of the lens must thus increase. This is brought about in the following way: the ring muscle contracts, with the result that the cross section of the ring becomes smaller. The tension in the ligaments decreases and the curvature of the lens becomes greater. The refracting power thus changes to adjust to the distance of the object, i.e. the eye accommodates. In this way it is ensured that the image is always on the retina, irrespective of the distance of the object from the eye.

Faulty image formation in the eye

The normal eye, the focal point of which is situated at the retina, is called an *emmetropic* eye (FIG. 14.16a). When the eye ball is flatter than normal or the refractive power of the lens is small, the parallel rays reaching the eye converge behind the retina (FIG. 14.16b). This condition is called *hypermetropia* or *far-sightedness* because the person sees objects that are far from him well, but those near to him only badly or not at all. The reason for this is that the rays coming from afar are refracted sufficiently to form an image on the retina, while rays coming from nearby and which are slightly divergent are not refracted sufficiently, and consequently the image is formed behind the retina. The condition can be corrected by using biconvex glasses.

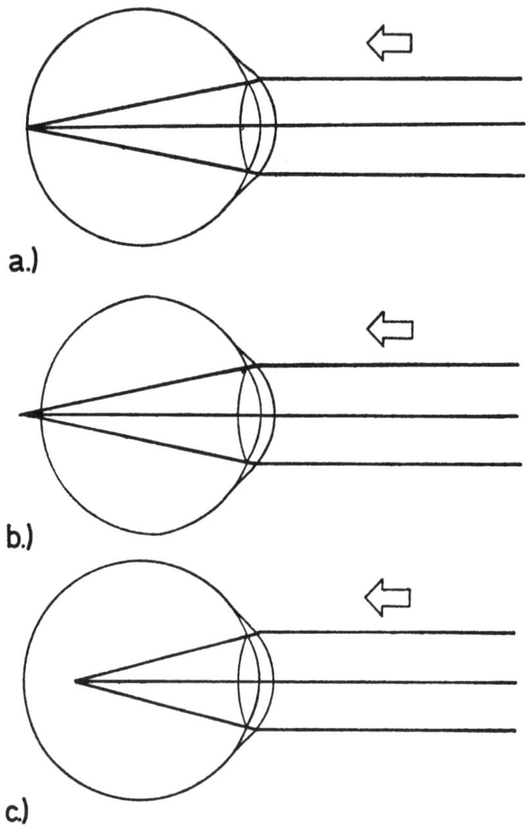

FIG. 14.16 (*a*) emmetropic eye, (*b*) hypermetropic eye, (*c*) myopic eye.

When the curvature of the lens is too great or the eyeball is rounder than normal (FIG. 14.16c and 14.17a), parallel rays coming from objects further than 6 metres converge in front of the retina. The result is an indistinct image. The condition is called *myopia*, or *short-sightedness*, since only objects close to the eye can be seen distinctly. The abnormal refracting power of the lens of the eye can be corrected by using a biconcave lens (Fig. 14.17b).

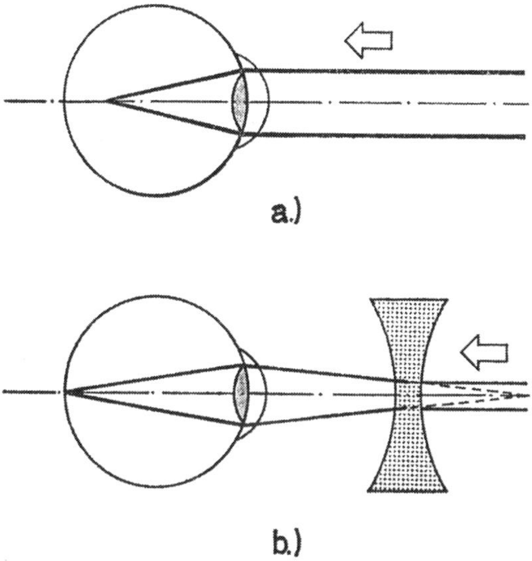

Fig. 14.17 (*a*) A myopic eye concentrates parallel rays in front of the retina. (*b*) The condition can be corrected by a biconcave lens.

As a person becomes older the accommodation mechanism of the eye loses its efficiency. This happens because the lens of the eye becomes less elastic and it can no longer adjust easily to the required changes. In this case objects far away can be seen easily but those that are close cannot be seen clearly. The condition is known as *presbyopia* and it can be corrected with suitable biconvex lenses.

CHAPTER 15

Radioactivity

Introduction

As mentioned earlier (Chapter 4), the difference between the isotopes of an element lies in the number of neutrons present in the nucleus of the element, and therefore also in the atomic mass of the element. All isotopes of the same element, however, *have the same number of protons*, and therefore the same atomic number. Certain isotopes are unstable, i.e., they disintegrate easily to form new elements. During the distintegration process, radiation is emitted. The rays have great penetrating power and are known as radioactive waves. Radioactive isotopes present in nature, such as radium and uranium, are called *natural radioactive isotopes.* Many of the radioactive isotopes, such as cobalt 60, iodine 131, chromium 51 and iron 59 are *produced artificially.* These artificially-produced radioactive isotopes are all used in medicine either for therapeutic or diagnostic purposes.
The radiation emitted by radioactive isotopes, has four properties which make the isotopes useful in medicine:

A. the radiation penetrates through living tissues with ease;

B. it *can be measured* with special instruments even when the radioactivity is minimal;

C. in small quantities it is *harmless*;

D. if a large quantity of the radioactive material is present the radiation produced will however destroy *living tissue.*

The first three properties make it possible to use radioactive materials for diagnostic purposes. The first and the fourth properties make the use of radioactive isotopes for therapeutic and sterilisation purposes possible.

Principles of radioactivity

When an unstable isotope disintegrates to form a new isotope, the *radiation that is emitted is characteristic of the disintegration process.* For example, when isotope A disintegrates to form isotope B, the type of radiation, the strength of the radiation and the rate at which the radiation is released are constant for the disintegration process in question, and are completely different from the type, strength and rate of radiation emitted when an isotope C disintegrates to form isotope D.

1. *Type of rays*

There are three types of radiation that can be emitted by radioactive isotopes, namely alpha (α), beta (β) and gamma (γ) rays. A certain isotope can emit one or more of them. The first two types actually consist of fast-moving particles, while the gamma rays are electromagnetic rays (FIG. 15.1). The particles of alpha rays consist of two protons and two neutrons, like the nucleus of a helium atom (Chapter 4). In the case of beta rays, the particles are electrons.

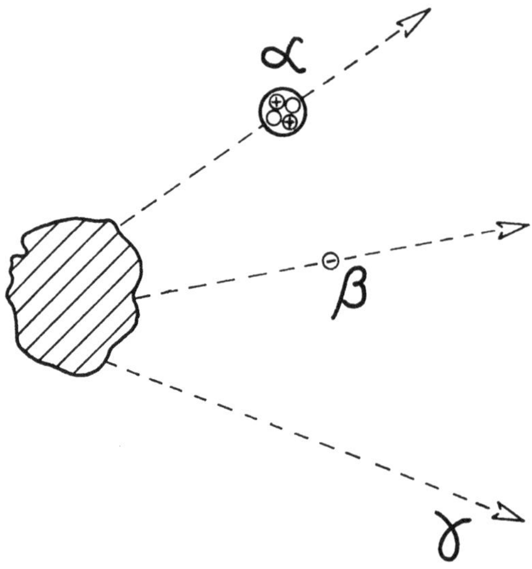

FIG. 15.1 A radioactive isotope emits three types of radiation. The α rays are moving particles consisting of two protons and two neutrons; β rays are moving electrons and γ rays are electromagnetic rays. A specific isotope may emit one or more of these rays.

2. *Intensity of rays*

The different radiations do not all have the same penetrating power, i.e., energy or intensity. Alpha and beta rays penetrate tissues only to a slight extent and can be blocked by a thin layer of metal (FIG. 15.2). Alpha particles can be stopped by a thin sheet of aluminium, having the thickness of the page on which this text is printed, and even in air they move as far as 30 to 80 mm only. Beta particles can be stopped by an aluminium sheet having a thickness of a few millimetres. Gamma radiation, on the other hand, is only absorbed by a very thick sheet of lead. The intensity of the radiation does not only vary with the type of emission, but radiations of the same type obtained from different disintegration processes may also have different intensities.

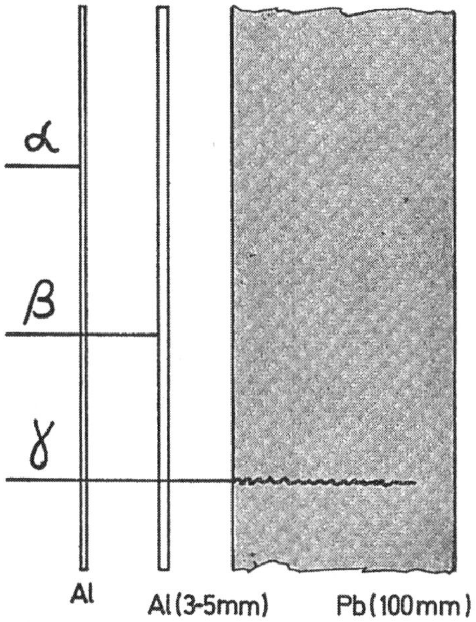

FIG. 15.2 The different radiations are absorbed to different extents. γ particles are absorbed by an aluminium sheet as thick as a sheet of writing paper. β rays by an aluminium sheet only a few mm thick; γ rays by a lead sheet of e.g. 100 mm thickness.

3. *Rate of disintegration. Half-life.*

The rate of disintegration is also characteristic of a certain disintegration process. Each radioactive isotope disintegrates in such a way that the percentage of the total atoms which disintegrate in unit time is constant for the certain isotope. The time it takes for half of the material to disintegrate is constant (FIG. 15.3). If, for example, we begin with 1 gram radium, 1 620 years will pass before $\frac{1}{2}$ gram disintegrates, i.e., has lost its radioactivity. After a further 1 620 years there will be $\frac{1}{4}$ gram radium left and after a further 1 620 years $\frac{1}{8}$ gram, and so on. A few further simple calculations will show that although in each set period of time (in this case 1 620 years) the radioactivity diminishes by half, it can never reach an arithmetical zero. It merely decreases to undetectable

Element	*Symbol*	*Radio-active isotope*	*Atomic mass*	*Atomic number*	*Rays emitted*	*Half life*
Calcium	Ca	^{47}Ca	47	20	$\beta-, \gamma$	4,7 days
Chromium	Cr	^{51}Cr	51	24	γ, K	27,8 days
Cobalt	Co	^{57}Co	57	27	γ, K	270 days
		^{58}Co	58	27	B+, γ	71 days
		^{60}Co	60	27	$\beta+, \gamma$	5,27 years
Gallium	Ga	^{67}Ga	67	31	γ	78 hours
Gold	Au	^{198}Au	198	79	$\beta-, \gamma$	2,7 days
Indium	In	^{113}mIn	113	49	γ	1,7 hours
Iodine	I	^{123}I	123	53	γ	13 hours
		^{125}I	125	53	γ	57,4 days
		^{131}I	131	53	$\beta-, \gamma$	8,05 days
Iron	Fe	^{52}Fe	52	26	$\beta+, \gamma$	8,2 hours
		^{59}Fe	59	26	$\beta-, \gamma$	45 days
Mercury	Hg	^{197}Hg	197	80	γ, K	65 hours
		^{203}Hg	203	80	$\beta-, \gamma$	47 days
Phosphorous	P	^{32}P	32	15	$\beta-$	14,3 days
Potassium	K	^{42}K	42	19	$\beta-, \gamma$	12,5 hours
Radium	Ra	^{226}Ra	226	88	α (The products formed when radium disintegrates also emit γ rays)	1 620 years
Sodium	Na	^{22}Na	22	11	$\beta+, \gamma$	2,58 years
		^{24}Na	24	11	$\beta-, \gamma$	15 hours
Strontium	Sr	^{85}Sr	85	38	γ	65 days
Technetium	Tc	^{99}mTe	99	43	γ	6,0 hours
Ytterbium	Yb	^{169}Yb	169	70	γ, K	32 days

TABLE 15.1 Some of the radioactive isotopes that are often used for diagnostic or therapeutic purposes. The data presented in the table have been considerably simplified to suit the purposes of this book.

K, e.g., ^{51}Cr: This indicates that the nucleus of an atom has absorbed one of its orbiting electrons (nuclear transformation or *K*-capture). The excess energy is shed by the nucleus in the form of X-radiation.

m, e.g., ^{99}mTc: Nuclear isomers of elements sometimes occur and are generally produced artificially. A nuclear isomer possesses the same atomic mass and atomic number as other isotopes of the element, but differs from these in its radioactive properties. The letter m, indicates "metastable", which means that although the nucleus of the atom is not in its most stable orm, it is nevertheless in a form which is able to lead a significantly independant existence. The metastable form of an element is indicated by using the letter m immediately after the figure indicating its atomic mass. Metastable atoms decay by shedding energy, for instance in the form of γ-radiation.

$\beta-$ and $\beta+$ particles: The signs — and + indicate negatively and positively charged β-particles which are encountered during the decay of some radioactive atoms. They are in fact two different forms of "electrons". Electrons are generally regarded as negatively charged atomic particles, but in nuclear physics the designation electron is often used to cover both negatively charged particles (negatrons) and positively charged particles (positrons). The — and + signs actually refer to the energy level of the particles concerned rather than to their electrical charge. Unlike electrons, positrons ($\beta+$ particles) are not normal constituents of matter. There is little difference in mass between positrons and electrons, but positrons have a very short half-life ($+10^{-9}$ seconds) since they disappear on collision with electrons which abound in their surroundings. The product resulting from the fusion of $\beta+$ and $\beta-$ particles is not a third type of particle but rather an energy emission of a type similar to γ-radiation.

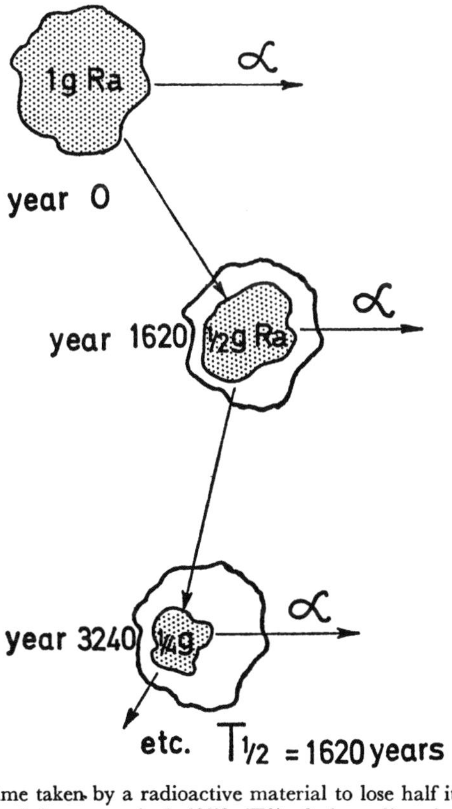

FIG. 15.3 The time taken by a radioactive material to lose half its radioactivity is constant. This time is known as the half-life (T½) of the radioactive material. When radium disintegrates, a whole range of breakdown products is produced that are radioactive themselves, and these can be used therapeutically. The most important products of decay together with the type of radiation emitted, shown in parenthesis, are the following in the case of radium, Ra.

Radium (α) → Radon (α) → Radium A (α, β) →
Radium B (β, γ) → Radium C (β, γ) → Radium C′ (α) →
Radium C″ (β, γ) → Radium D (β, γ) → Radium E (β, α) →
Polonium (α) → Lead (stable).

limits. During the process of disintegration, which it must be stressed is merely a loss of radioactive properties, the element is transferred

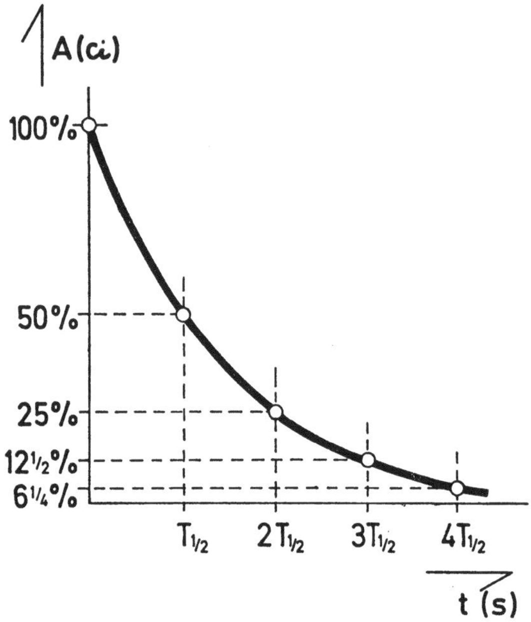

FIG. 15.4 Graphic representation of radioactive disintegration. A = activity, t = time.

into another stable element. Radium, for instance, is transformed into lead. Disintegration must be regarded as the transformation of an element into another one rather than the complete disruption or disappearance of the elemental matter. This transformation, as we have seen, happens to depend upon the emission of energy in the form of various radiations. Another useful word for disintegration is decay or more properly, radioactive decay. The constant time that it takes for the half of the isotope to undergo radioactive decay, is known as the half-life ($T_{\frac{1}{2}}$) of the isotope (FIG. 15.4). Some radioactive isotopes have a very long half-life, sometimes thousands of years, like radium. Other radioactive isotopes have a half-life of days, minutes or even seconds.

For each atom of an isotope that disintegrates, a certain quantity (quantum) of radiation is formed. It is therefore obvious that isotopes which disintegrate fast have a short half-life and those that disintegrate slowly have a long half-life. The activity of a radio-active substance, or in other words, the radiation emitted by the substance, is directly proportional to the number of atoms that disintegrate. The unit of radioactivity is the curie (abbreviation: Ci). One curie is equal to 3,68 $\times$ 10^{10} atoms disintegrating per second. This number is approximately equal to the number of atoms disintegrating per second in 1 gram of radium. (Radium was the first radioactive isotope discovered by Pierre and Marie Curie).

Doses of radioactive substances used for diagnostic purposes in medicine are usually in the microcurie range, i.e., $3{,}68 \times 10^{10} \times 10^{-6} = 3{,}69 \times 10^{4}$ atoms disintegrating per second. The quantity of radiation emitted is, however, still sufficiently high to be picked up by special apparatus.

Diagnostic uses of radioactivity

Principles of diagnostic instruments. As mentioned previously, the properties of radiations which make their use possible in diagnosis are their penetrating power, their measurability and their safety when administered in small quantities. Furthermore, we know that *the chemical properties of all isotopes of an element are the same.* The reason for this is that the chemical properties are determined by the number of protons and electrons, which are the same for all isotopes of the same element. The human body like other biological systems, has the ability to use certain elements or compounds in a specific manner. For example, iodine is concentrated mainly in the thyroid or the kidney. The kidney excretes it, while the thyroid concentrates it for use in the production of the thyroid hormones. The use of radioactive iodine to investigate the functional status of the thyroid gland is based on the fact that the thyroid selectively concentrates iodine and does not distinguish between ordinary iodine and radioactive iodine, since chemically they do not differ. While the thyroid functions normally, its hormones are produced at a uniform rate, i.e., the gland concentrates iodine at a uniform rate and this iodine is distributed evenly throughout the gland tissue. If the gland produces too much or too little of these hormones, it will concentrate too much

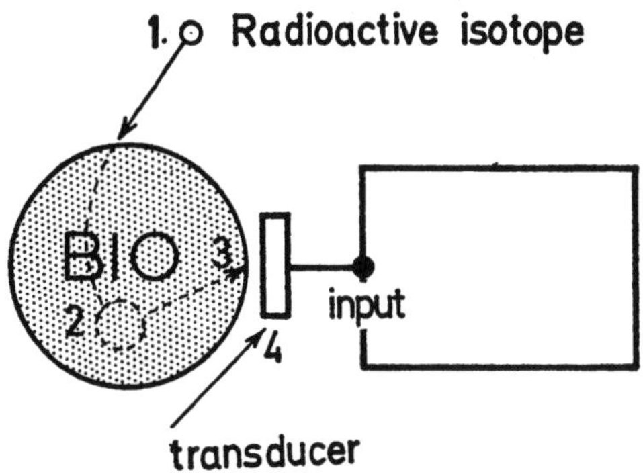

FIG. 15.5 Schematic representation of the principles on which diagnostic apparatus is based for use with radioactive substances (see text).

or too little iodine, respectively. The gland will also concentrate iodine unevenly if different parts of the gland are not equally active, as for instance in cases of cancer of the gland. With the aid of radioactive iodine isotopes, the rate of iodine concentration and distribution in the gland can be determined. Other radioactive isotopes are used to investigate the kidneys, liver, gallbladder, pancreas, lungs, heart, uterus, skeletal system, blood, brain and so forth.

The procedure of a diagnostic investigation (FIG. 15.5)

1. A known small quantity (a few microcurie) of the relevent radioactive isotope is mixed with the normal element. We say that the

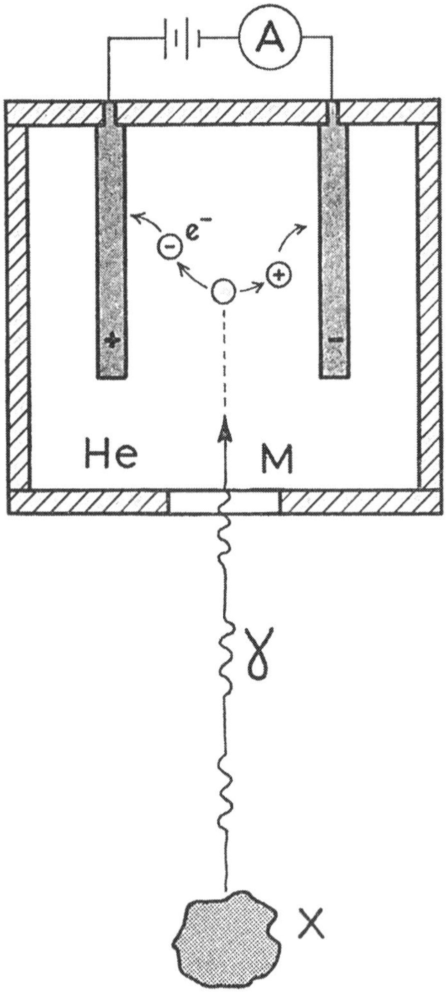

FIG. 15.6 Schematic representation of the principles according to which a Geiger counter operates.

element is "labelled". The labelled substance is given to the patient by mouth or intravenously.

2. The labelled substance follows the same physiological route in the body as the unlabelled substance and is distributed throughout the body or becomes concentrated in a tissue or is excreted by a particular organ depending on its nature.

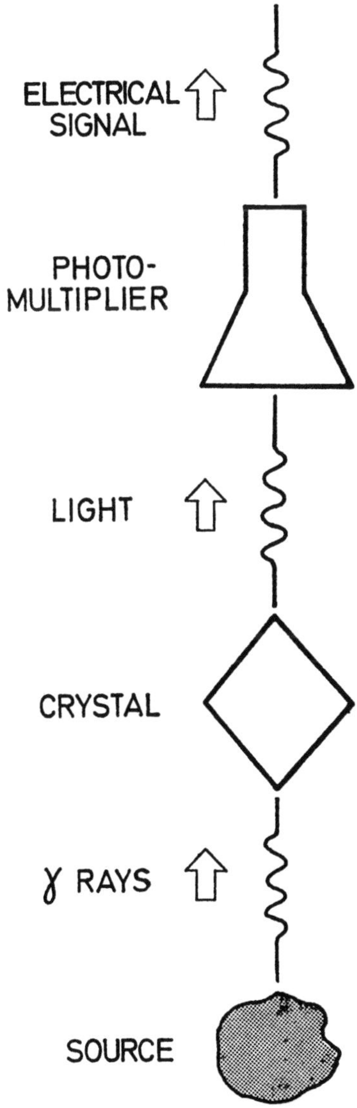

Fig. 15.7 Principles of radioactive transducers (see text).

3. The radioactive emissions from the isotope penetrate the tissues and leave the body.
4. The radiations are picked up by a transducer which converts them into electrical impulses. An electrical apparatus amplifies and registers the impulses. The diagnostic apparatus is therefore a secondary input device (Chapter 7). The registration, in normal cases, has a specific form, and deviations from this can be used to reveal certain functional and structural disturbances.

Technical information on diagnostic apparatus

a) Direct registration of radioactive rays. The simplest way to

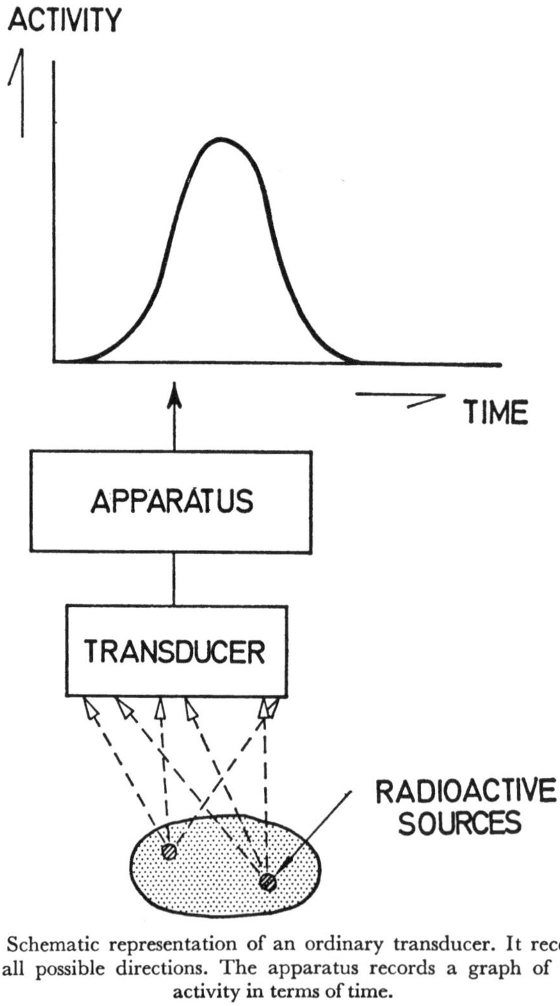

FIG. 15.8 Schematic representation of an ordinary transducer. It receives radiation from all possible directions. The apparatus records a graph of total radioactivity in terms of time.

register radioactivity is by means of a *photographic film.* Since gamma rays and X-rays are electromagnetic radiations, like light is, they will affect photographic paper in the same manner as done by light. Photographic film is also affected by alpha and beta particles. This finds an important use in the wearing of a small photographic film by persons who work with radioactive materials and X-rays. The film registers the quantity of radiation to which the person is exposed and is therefore an important safety device in the detection of excessive irradiation.

(*b*) Transducers. The simplest transducer of radioactivity is the *Geiger-Müller counter* (FIG. 15.6). It works according to the principle that certain gases ionise under the influence of radioactive rays. The greater the intensity of the radiation, the greater is the ionisation in the gas. A gas containing ions is a good conductor of electricity, and the more ions present, the better will electricity be conducted. The gas that is generally used in the Geiger-Müller counter is helium. The gas is held in a container provided with a special window, through which radiation can enter. Two metal electrodes which make contact with the gas are connected to a battery and thus provide a potential difference. Nonionized helium does not conduct electricity and the meter that is connected in the

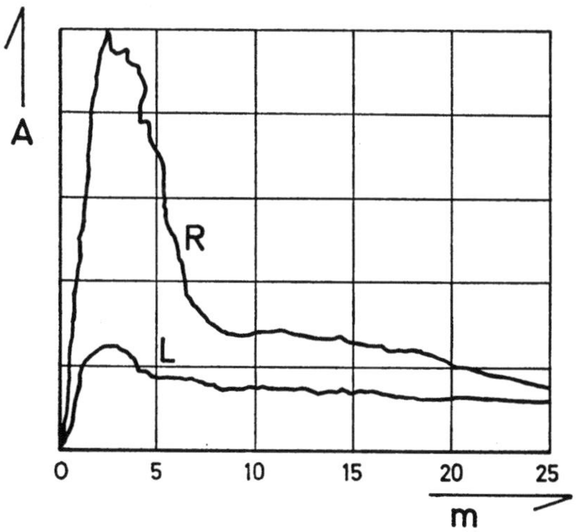

FIG. 15.9 Renogram. In the investigation, a radioactive isotope is injected intravenously and the passage of the isotope through the kidney is registered graphically. R = right kidney (normal); L = left kidney (abnormal); A = radioactivity; m = time in minutes.

circuit will register zero. The radiation from the isotope concerned causes the flow of a current in the system, the intensity of which is measured on the meter. An input device can be used instead of the meter to register the radioactivity. One method is to connect a loudspeaker to the system. It must be remembered that the radiation arrives in quanta, and consequently the current reaches the apparatus in bursts. This produces a sound resembling that produced by a machine gun, in the loudspeaker.

The transducer that is most commonly used comprises a *scintillation crystal and a photomultiplier* (FIG. 15.7). A scintillation crystal emits light rays when radiation falls on it. The photomultiplier converts the light rays into an electrical potential that can be fed into the counting apparatus.

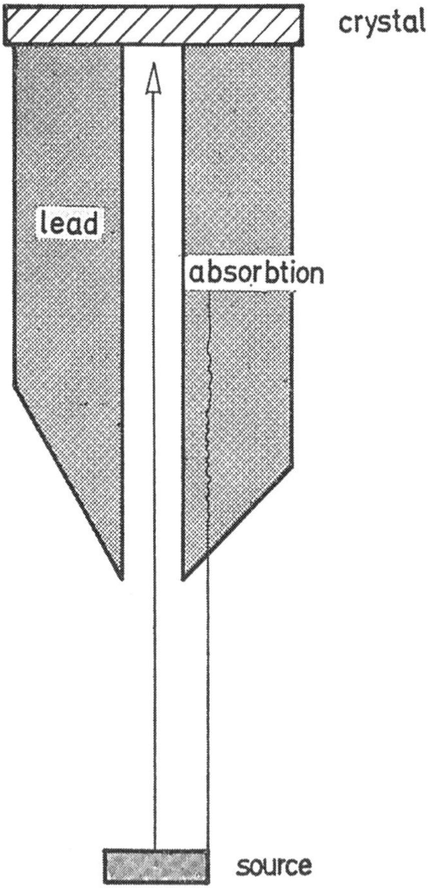

FIG. 15.10 Schematic representation of a collimator housing a crystal. The collimator allows only rays from directly beneath the crystal to pass through.

There are various types of transducers available. If the distribution of the measured activity is not of importance, a single large crystal can be used (FIG. 15.8). The crystal receives radiations from different centres of radioactivity distributed throughout the organ concerned and therefore measures the total activity of these centres rather than their distribution. This kind of crystal is used, for example, to investigate kidney function. In this case one of the normal constituents of blood is labelled with radioactive material. As the radioactive substance moves through the kidneys, the activity measured by a crystal placed immediately above the kidneys increases first and then decreases again. The change in

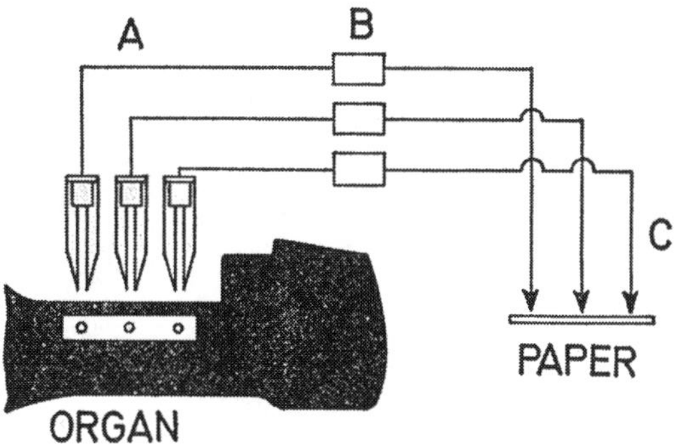

FIG. 15.11 Principles of the scintillation camera. A = transducers, B = amplifiers, C = recorder.

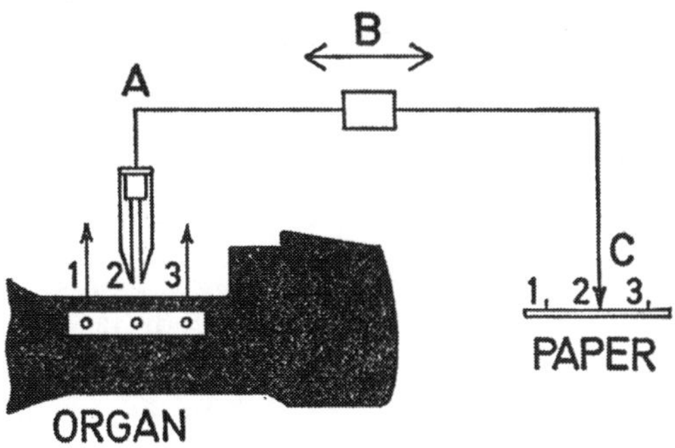

FIG. 15.12 Principles of the scintillation scanner. A = moving transducer, B = amplifier, C = moving recorder.

activity is recorded in the form of a graph on moving paper. Under normal circumstances the graph has a definite shape (FIG. 15.9), which, however, changes if the kidneys do not function properly. The graph can thus be used for diagnostic purposes.

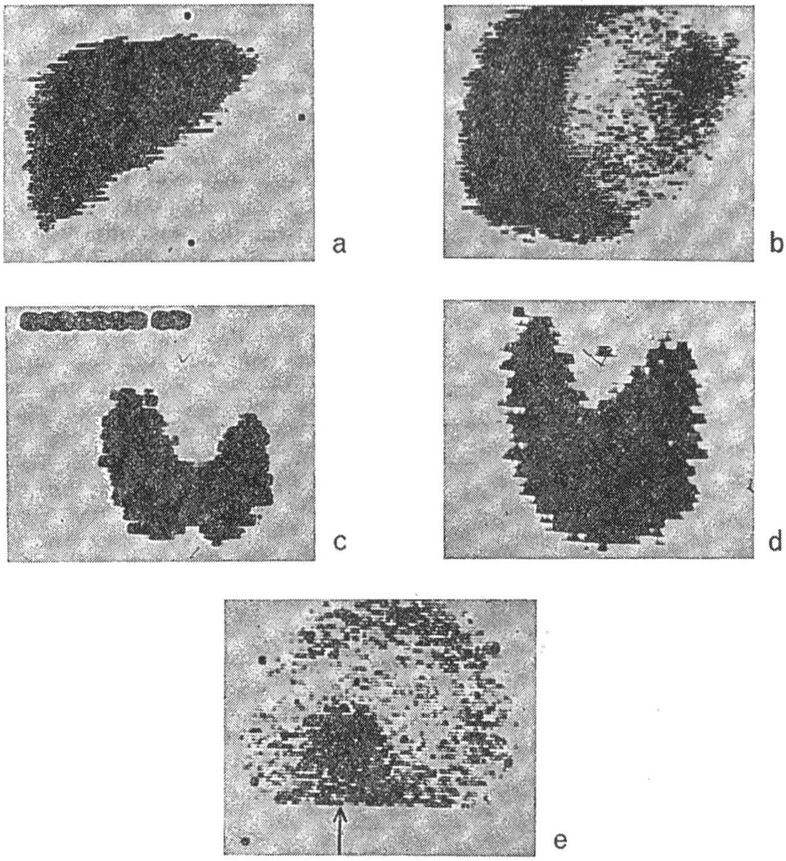

FIG. 15.13 Scintigrams of a few organs (*a*) normal liver, (*b*) enlarged liver with degeneration of part of the liver, (*c*) normal thyroid gland, (*d*) enlarged over-active thyroid gland, (*e*) brain tumour indicated by arrow.

If only small areas in an organ are to be studied, a small crystal placed in a *collimator* can be used. The collimator is a tube which houses the crystal and is constructed in such a way that the crystal receives only the rays that are produced directly beneath it (FIG. 15.10).

Should one wish to examine every small portion of an organ separately, this can be done by placing a number of the above-mentioned transducers next to each other (*scintillation camera*, FIG. 15.11), or one transducer can be moved from place to place (*scintillation scanner*, FIG. 15.12). In both cases the activity is registered in the form of points (FIG. 15.13). The registrations are called scintiphotos and scintigrams, respectively.

Therapeutic uses of radioactivity

Because intense radioactive irradiation destroys living tissue, it can be used to treat tumours and destroy malignant cells. Treatment is performed according to the following three methods.

1. If tumours or cancerous cells have a special affinity (preference) for some radioactive substance it can be injected into the bloodstream or be given by mouth and is then selectively absorbed or concentrated by the tumour cells which consequently receive radiation from within themselves. An example of this kind of treatment is the injection of phosphorus 32 (^{32}P) to suppress excessive red blood cell production by tumour cells in the red bone marrow. Another example is the administration of radioactive iodine in the treatment of a hyperactive thyroid gland or in cases of cancer of the thyroid gland. Unfortunately only a small percentage of the cancers of the thyroid are able to concentrate iodine selectively.

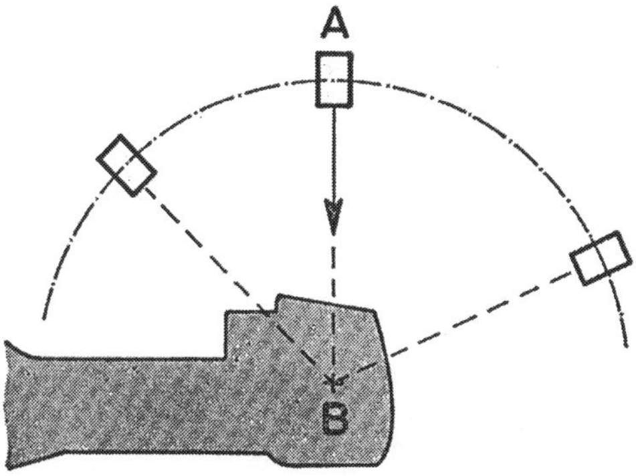

Fig. 15.14 Schematic representation of the principles on which teletherapy is based. A = moving radioactive source, B = the area being irradiated.

2. The radioactive substance can be implanted in the form of *wire* or *pellets* directly into the area of the tumour. Examples of such substances are tantalum 182, radium and radioactive gold.
3. A strong radioactive source can be used to irradiate the malignant cells (Fig. 15.14). To avoid the destruction of normal cells, the source is moved in a circle in such a way that the point to be irradiated is at its centre. All the rays pass through this central point, but only a few pass through other points in the body. This type of treatment is called *teletherapy*, and an example of it is the *cobalt-source* which uses the isotope cobalt 60.

Beta particles can also be produced by means other than radioactive decay. These particles are actually electrons moving at a very high speed and having therefore a high kinetic energy (Chapter 3). Apparatus such as the *betatron*, *synchrotron* and the *linear accelerator* are used to produce high energy beta radiation which can also be used therapeutically. The destructive effect of gamma radiation is also used to sterilise air, containers, rooms and solutions, since the radiation destroys micro-organisms.

Precautions concerning the use of radioactive isotopes

Because of the dangers connected with radiation, persons working with radioactive materials must take special precautions to protect patients and themselves from the radiation concerned. In accordance with what has been said before, there are two types of sources of radioactive radiation, namely:

(1) external sources — apparatus providing mainly gamma rays,

(2) internal sources — radioactive material that has entered the body. This type of contamination is very dangerous because the source is in very close contact with the tissues and a small amount can therefore cause great damage. Also not all types of radioactive materials which enter the body disintegrate rapidly or are rapidly eliminated.

The following are a few of the precautions recommended.

1. A protective apron should be used when working with radiation.
2. Forceps and gloves should be used when handling radioactive isotopes.
3. The time of exposure to radioactive material or radiation should be kept to a minimum.
4. Avoid spilling radioactive material.
5. The handling of patients and bed linen should be carried out strictly according to the prescribed rules.
6. Every person working with radioactive materials must wear a film badge that should be developed regularly. In this way the person knows exactly how much radiation he has been exposed to.

CHAPTER 16

Sound and Hearing

Introduction

Sound is the most important form of communication between people, *and it can be defined as the phenomenon which stimulates the auditory receptors of people with normal hearing.* In air, sound consists of repeated air pressure changes, i.e., alternatively increasing and decreasing pressures, the frequency of which varies from 25 to 18 000 hertz. These pressure changes cause the elastic ear-drum, which is actually a part of the auditory receptor, to oscillate; in this way the receptors become stimulated. As a result of the stimulation, impulses are produced in the receptors which are transmitted to particular areas in the brain, producing particular sensations.

The vocal cords of human beings, loudspeakers and musical instruments are specially constructed to produce sound, i.e., oscillations of the appropriate frequencies. The volume of the layer of air in contact with such an oscillating structure changes, and according to the gas laws (Chapter 9), this change in volume causes pressure changes. The volume changes, together with the accompanying pressure changes, are transmitted in all directions to successive layers of air. The pressure changes at any point in the vicinity of the sound source can be plotted graphically against time. The graph shows a sine wave pattern (FIG. 16.1), and hence one speaks of sound waves.

The difference between electromagnetic waves and sound waves can be summarised as follows: (1) in the case of electromagnetic waves, the electrical and magnetic forces oscillate on either side of direction of propagation of the waves, whilst in sound waves the pressure changes are in the *direction of the movement;* (2) sound, in contrast to electromagnetic waves, cannot be transmitted in a vacuum. On the other hand, the two types of waves are similar in that they are both transmitted in all directions, and as a result their intensity decreases with the square of the distance from the source (Chapter 13). The laws of reflection and refraction also apply to sound waves (Chapter 14).

Properties of sound

1. *Pitch.* The human ear can hear a whole range of frequencies, which are therefore called audible frequencies. The sensations in the

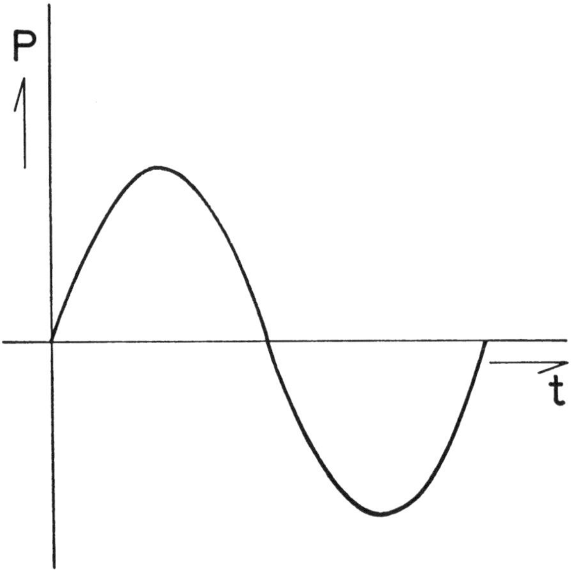

FIG. 16.1 Graphical representation of the pressure changes (P) versus time (t) during the conduction of sound.

brain caused by the various frequencies differ from each other. Slow frequencies are interpreted as low pitch and fast frequencies as high pitch.

2. *Music and Noise.* A pure tone consists of a single frequency. Sound, however, seldom exists as a pure tone, but is made up of different frequencies occurring at the same time. If the different frequencies comprising the sound are in a *simple ratio to each other* the result, as in music, is *pleasant.* If, however, the *frequencies are unrelated* the result is *noise.*

3. *Fundamental tone and Overtones.* Musical instruments deliver pleasant sounds produced by the vibrations of a string, a membrane or an air column. In the case of the piano, the best-known string instrument, the strings are struck by small wooden hammers covered with felt. A string of a certain length stretched to a certain tension can produce different frequencies, including a fundamental tone and overtones. If the string is struck at its centre, oscillation of the entire string occurs, and the fundamental tone is obtained (FIG. 16.2a). If the string is struck half-way between the centre and the end, the two halves of the string oscillate alternately, and the frequency of oscillation is twice that of the fundamental tone (FIG. 16.2b). A note

having a frequency of twice that of the fundamental tone is called the first overtone. If the string is hit at a point one sixth of its length from one end, the second overtone is obtained (FIG. 16.2c).

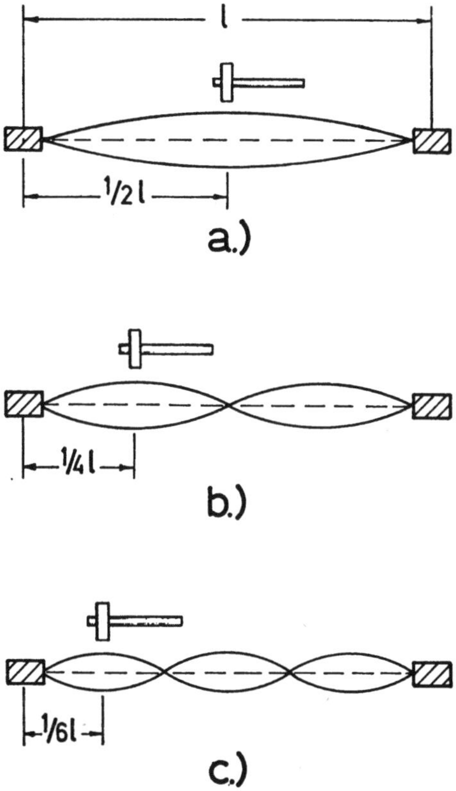

FIG. 16.2 When a taut string is tapped on at various points, it produces vibrations of different frequencies (see text).

4. *Quality of tone*. The sounds produced by the strings of the piano or another musical instrument consist of a mixture of fundamental tone and overtones. The fundamental tone is always the most important, since it determines the note, while the overtones contribute only a small percentage to the sound. The effect given by such a *mixture* is called the *quality of tone*. The quality varies from instrument to instrument since different instruments produce differing numbers of overtones. This is the reason why the same note produced by two musical instruments, for example a piano and a violin, sounds different. *Middle C*, for example, always has a fundamental note

formed by 256 vibrations per second; the overtones of different instruments provide further contributions of 512 (i.e., 2 × 256), 768 (i.e. 3 × 256) and still higher vibrations per second.
Sound produced by a string already containing a certain percentage of overtones obtains more overtones from the vibration of the framework or cabinet of the instrument. The dimensions of the instrument, therefore, contribute to the quality of tone. In man the principles of sound production are the same as for musical instruments. The note produced by the vocal cords causes vibration of structures in the mouth and nostrils. The quality of tone is, therefore, dependent upon the anatomical construction of the sound producing apparatus in man.

5. *Pleasant Sound and Harmony.* Music is pleasant or harmonious if the different notes that are emitted together or in close succession consist of frequencies related to each other in simple ratios. *Middle C* for example has a frequency of 256 hertz and *high C* 512 hertz. When the two are played together, they produce a pleasant effect because their frequencies are in the ratio of 2:1. This ratio is called an *octave.* Other simple ratios producing a pleasant effect are 2:3, 3:1, etc.

6. *Intensity* is determined by the amplitude of the vibrations. If the string is hit harder, the sound produced will be stronger because the stronger pressure changes cause stronger oscillations in the receptors of the ear. The intensity is usually measured in *decibels* (*dB*).

7. *Speed of Sound.* In air at 20°C sound is transmitted at a speed of 340 m/s. In water it moves four times as fast, and in iron 15 times as fast.

Production of sound

Anything that vibrates at an audible frequency produces sound. The best known ways of producing pure sound are by using strings or air columns. The vocal cords are in effect strings.

1. *Sound production by strings.* Musical instruments such as the piano, violin, cello and guitar are stringed instruments. The strings are activated in various ways; for example with a small hammer, with a bow or with the fingers. The pitch of the fundamental tone produced by a particular string is determined by the length and thickness of the string, and also by the tension in the string. That these properties do indeed play a role is evident from the way a violin or a guitar is played. (1) If the artist wishes to play a higher tone, he reduces the length of the string with his fingers. (2) If he wishes to go still higher, he uses one of the thinner strings. (3) If he requires a higher tone from a particular string, he stretches the string.

The tuning fork, used quite often in medicine to perform tests for hearing, is actually a firm string.

2. *Production of sound by vibrating air columns.* In wind instruments the air is blown through a mouthpiece into a pipe, and this causes vibration of the air present in the pipe. The mouthpiece may have a vibrating reed or a special opening through which air is forced. The cross-section and length of the pipe determines the frequencies of sound produced by such an instrument.

3. *Sound production by man.* The sounds emitted during singing or speech are produced in two stages. The first stage of speech is carried out by the vocal cords and is called phonation. This consists of the formation of the fundamental tone, together with contributions by overtones. The second stage of sound production is the changing of the quality of the sound by structures in the mouth and the nose.

The vocal cords consist of folds on both sides of the larynx situated in the air passage between the mouth and the lungs. As air moves through the larynx, the vocal cords vibrate, and this vibration is transmitted to the moving air. The position of the vocal cords and the degree of tension in them is controlled by the muscles of the larynx. The contraction of these muscles is under conscious nervous control from the brain. The vocal cords produce different frequencies according to their form and tension.

During the second stage of sound production which occurs when the air is moving through the mouth, the quality of sound is changed because some overtones become absorbed or smothered, while new ones are introduced. The organs that play a role in this respect are the lips, the tongue, the teeth and the soft palate. It is evident that sound production in human beings follows the same principles applicable to string and wind instruments.

Perception of sound by the ear

The function of the ear is to convert the pressure variations in the air into mechanical vibrations. These then produce electrical impulses in the auditory nerves of the ear which are then transmitted to the brain. The impulses cause special sound sensations in the area of the brain involved in sound perception. The ear can therefore be regarded as a transducer which changes the pressure variations into electrical signals. The ear consists of three compartments: the external ear, the middle ear and the internal ear. The external ear receives the vibrations of the air and conducts them via the auditory canal to the middle ear. A membrane, the ear-drum, separates the external and middle ear. When the air waves reach the ear-drum, they produce pressure changes, causing the latter to vibrate. The vibrations in the ear-drum are transmitted to the inner ear by means of the auditory bones.

The internal ear which contains a fluid, houses the receptors for hearing and body equilibrium. The auditory receptors are situated on a membrane consisting of about 24 000 parallel fibres (auditory strings). The lengths of the fibres vary and each fibre is connected to a fibre of the auditory nerve. Sound waves reaching the ear are transmitted to the auditory strings and cause them to vibrate at the frequency and amplitude of the original wave. The vibrations produce impulses in the auditory nerve which are conducted to a certain section of the brain, where they produce the required auditory sensations.

Hearing abnormalities

If some part of the hearing apparatus does not function properly, a person becomes partially or completely deaf. Two types of deafness exist: (1) *nerve deafness*, where the malfunction is in the internal ear or in the nerve itself, and (2) *conduction deafness*, where the fault is in the middle ear or in the ear-drum. In the second type of deafness, the mechanisms that change the pressure variations into mechanical variations are at fault. In this case oscillations can still be produced in the middle ear by conduction through the bones of the head. This latter principle is used to determine the type of deafness in a patient. A person with conduction deafness will not hear a vibrating tuning fork held next to his ear, but will *hear* the vibrations when the handle of the fork is pressed against his head behind the ear. In the case of nerve deafness, the patient will not hear the tuning fork in either of the above positions.

Hearing abnormalities can be assessed more accurately by means of an audiometer. The *audiometer* is a secondary diagnostic output apparatus that can produce pure tones of different frequencies. The amplitude of these can be read from a scale in decibels. The amplitude which can just be heard by a person with normal hearing is taken as the zero point. In the case of a patient who is partially deaf, the amplitude is increased until he is able to hear the sound. The amount of amplification indicates the degree of loss of hearing of the patient and it is usually expressed in decibels. The same examination is repeated at various frequencies with air conduction and bone conduction. In the case of air conduction, the sound is transmitted by an earphone; in the case of bone conduction, by an oscillating object pressed against the head. With the aid of this apparatus one is able to obtain a graph which indicates the sharpness of hearing in relationship to the frequency of oscillations. The registration is called an *audiogram* (FIG. 16.3).

Partial nerve or conduction deafness can be corrected with *hearing aids*. With partial nerve deafness the hearing apparatus still functions, but inefficiently. This condition is found in elderly people, and can be corrected simply by amplifying the sound with a hearing aid before it reaches the ear drum. The hearing aid consists of a micro-

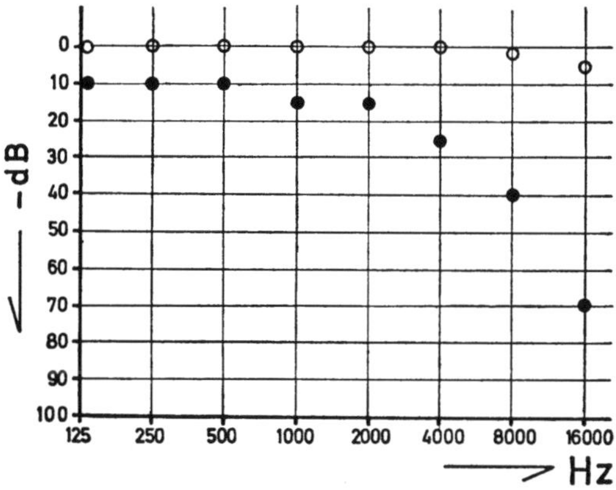

FIG. 16.3 An audiogram is a graphical representation of the acuteness of hearing in decibels (dB) in relation to the frequency (Hz) of sound.

phone, an amplifier and a loudspeaker placed in the external ear. In total conduction deafness, the sound waves must be transmitted through the bones behind the ear by a special vibrating object. A hearing aid is a secondary therapeutic output apparatus.

Sound amplifiers

When the vibrations produced by an object are so weak that their intensity is insufficient to initiate impulses in the sound receptors, sound amplifiers can be employed to produce the required intensity. This is the principle used to amplify heart sounds and heart murmurs. The heart *sounds* are produced partially by turbulent blood flow, partially by vibration of the heart valves and partially by vibration of the heart muscle. The oscillations are transmitted to the chest wall and can be registered there. The simplest way to transmit the vibrations to the ear of an observer is by means of a stethoscope. The stethoscope consists basically of an elastic tube filled with air. When one end of the tube is placed on to the chest wall and the other on to the ear of the observer, the oscillations produced in the chest wall cause oscillations in the column of air in the pipe, which are then transmitted to the ear of the observer. Electronic stethoscopes can also be used to amplify sounds. These are secondary input apparatus that amplify the vibrations picked up by means of a microphone. The amplified signals can be fed into a loud-speaker or they can be recorded by means of an oscilloscope or a pen recorder. In the latter case the apparatus is called a *phonocardiograph* and the registration a *phonocardiogram.*

Ultrasonics

According to the foregoing discussion, sound is caused by mechanical vibrations of materials and is responsible for the production of specific sensations in the brain. Although there exists a frequency limit for audible sound, oscillations with still higher frequencies do exist. Certain species of animals can perceive frequencies of 20 000 hertz or more. Oscillation frequencies exceeding the audible range of a person with normal hearing are called ultrasonic frequencies (ultra = above). Ultrasonic vibrations can be produced by a special crystal through which an alternating current is passed. The oscillations can again be picked up by a similar crystal connected to an electronic apparatus. The ultrasonic vibrations can be used for therapeutic, surgical and diagnostic purposes.

1. *Therapeutic uses.* These are based on the fact that if the amplitude of the oscillations is fairly strong, they will cause vibrations in the tissues accompanied by the production of heat. The heat so produced assists in the healing of inflamed joints and injured muscles. Therapeutic ultrasonic apparatuses are secondary output devices, the crystal playing the role of an output transducer.
2. *Surgical use.* If the frequency and the amplitude of the oscillations is sufficiently high, they will destroy living tissue. This principle is used for the surgical treatment of certain diseases.
3. *Diagnostic use.* When ultrasonic oscillations move from one medium to another, some pass through, while the rest are reflected. This principle is used to search the brain, liver and the heart for anatomical aberrations. The reflected ultrasonic vibrations are picked up by the input apparatus and are studied by means of an oscilloscope. The presence of an abnormality influences the vibration pattern and assists the doctor in making his diagnosis. The methods used to examine the brain and the heart are called *echoencephalography* and *echocardiography*, respectively. The apparatus used for such investigation is also called a *diasonograph.* Ultrasonic apparatus is also used in obstetrics and dentistry. Diagnostic ultrasonic apparatuses are secondary output-input apparatuses (Chapter 7). The destructive effect of oscillations on bacteria is used to sterilise surgical instruments.

SECTION V

MECHANICS

CHAPTER 17

Machines

Introduction

In Chapter 3 it was shown that forces produce movement if movement is allowed to take place. The skeletal muscles produce forces enabling man to move his body or the different parts of his body and with the aid of the latter to move other objects. The forces that are developed in the muscle are electrical (Chapter 3). As a result of these forces, the muscle filaments slide over each other causing the fibre, and therefore the muscle, to contract. The force developing in a muscle is the sum of the forces produced by the large number of fibres constituting the muscle. The effect of the force is limited, because:

(1) the *direction* in which movement can take place is limited to a single one, namely the direction of the muscle fibres,

(2) the *magnitude* of movement is limited to the extent to which the muscle is able to shorten, i.e., only a few centimetres even in the largest muscle.

It is general knowledge that in spite of the above-mentioned limitations the human body is able to produce a large variety of movements. This is made possible by altering the magnitude, direction and the extent of movement of forces originally produced by the muscles. A mechanism that can change these three parameters is called a machine. There are three simple machines: the *lever*, the *pulley* and the *incline*, and all other machines are composed of these three simple machines. The body consists of simple machines, using essentially the lever and the pulley composed of bones, tendons and joints. The forces developed by the human body are not always sufficient to perform tasks desired by man, and these forces must therefore be altered further. This is done by artificial machines. Some of which are used in medical practice and will be discussed later in this chapter.

Simple machines of the body

In FIG. 17.1 a schematic representation of change in direction of a force is shown. The large muscle of the upper leg produces a force

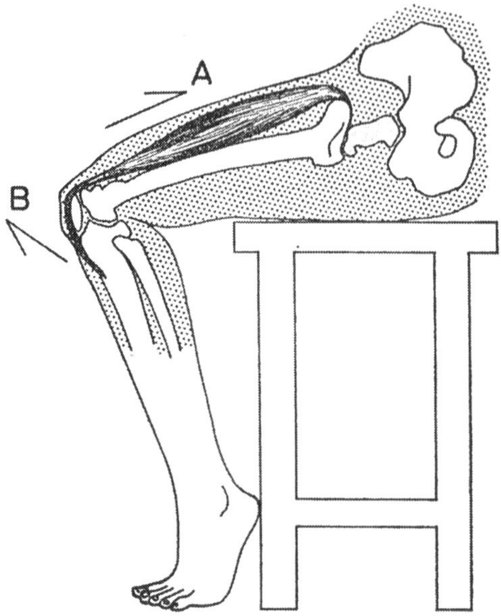

FIG. 17.1 The muscle produces a force in the direction A, but the movement takes place in direction B.

which is used to lift the lower leg. This is achieved as follows: the lower end of the muscle is joined to a tendon, connected at its opposite end to the bones of the lower limbs. Where the tendon passes over the knee joint, it widens to cover the knee-cap partially. The deep surface of the knee-cap is covered with cartilage and there is also a small amount of fluid between the deep surface of the knee-cap and the outer surface of the bone on which it slides. The smooth surface and the fluid allow movement without friction. The tendon, together with the base on which it is able to slide without friction, may be regarded as a machine, since it changes the direction of force.

We can construct a similar machine by allowing a string to move across a fulcrum (FIG. 17.2a). With such a machine the direction of the string, and thereby the direction of the force, is altered. In this way, for example, a great mass can be lifted against the gravitational force by another force acting in any other suitable direction. An important point is that the string must move over the support with ease and without friction, since the latter hinders movement.

This problem can be solved by passing the string over a groove in a circular disc rotating around a central axis (FIG. 17.2b). The fric-

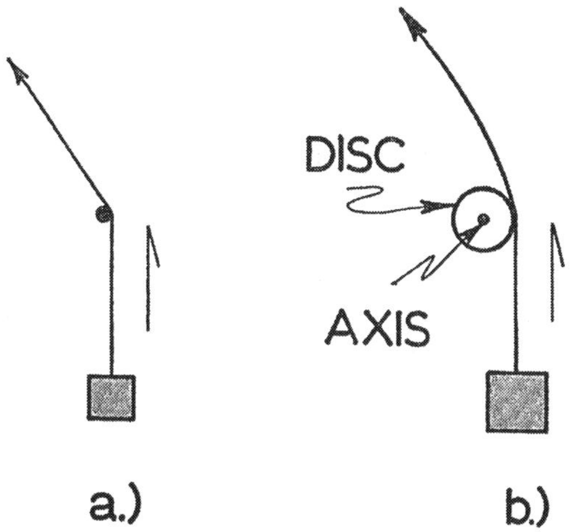

Fig. 17.2 Schematic representation of the principles of a pulley (see text).

tion between the axis and the disc can be reduced to a minimum by lubricants. This type of machine is called a pulley.

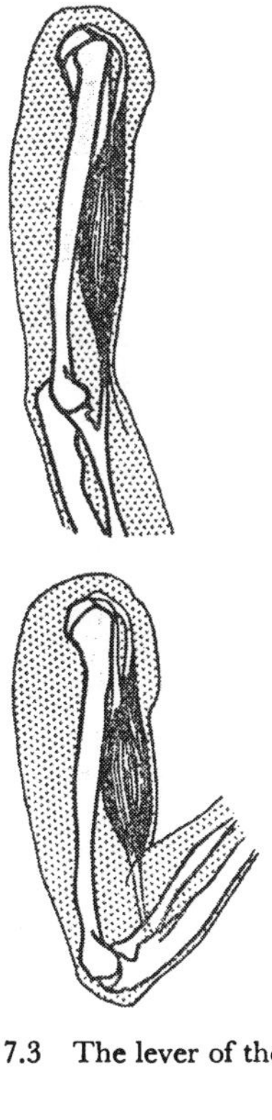

Fig. 17.3 The lever of the arm.

Another example of a simple machine is to be found in the arm (FIG. 17.3). The muscles of the upper arm are used to lift the lower arm together with the objects that might be in the hand of the person. If a person is standing, he can lift an object from a level corresponding to the midpoint of his upper leg to the shoulder, without moving the upper arm. This is achieved by contraction of the biceps. An interesting aspect of this is that although the muscle only shortens a few centimetres, the final extent of movement is about one metre. The machine, which in this case comprises the elbow serving as fulcrum, and the bones of the lower arm, changes the extent of movement. This type of machine is called a lever.

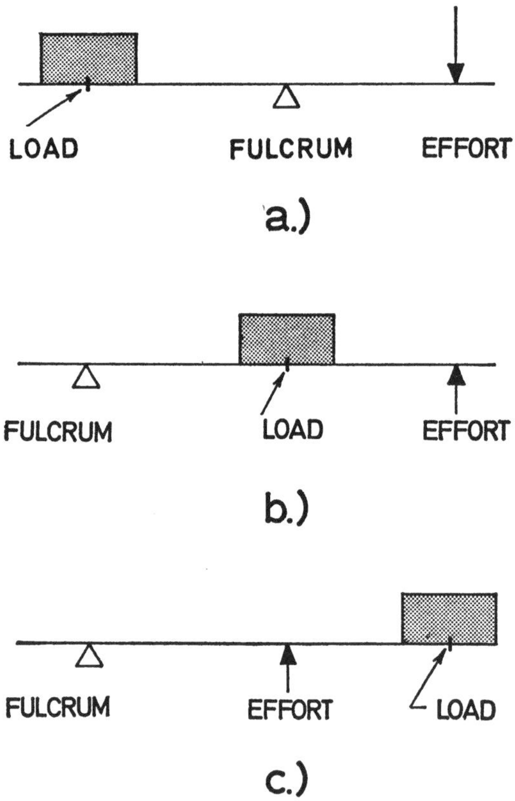

FIG. 17.4 The different types of levers: (a) first, (b) second and (c) third class levers.

Levers

All solid bodies that can rotate about an *axis* or a *fulcrum* are called levers. Besides the fulcrum there are two other points on a lever that are important, namely the point of application of the load and the point where the effort is exerted. Levers can be divided into three separate groups according to the relative positions of these three points.

First class levers. In this case the fulcrum lies between the points of application of the load and effort (FIG. 17a).

Second class levers. Here the load is applied between the fulcrum and the effort (FIG. 17.4b).

Third class levers. In this case the effort is placed between the load and the fulcrum, as is the case with the lower arm rotating about the elbow (FIG. 17.4c).

In order to understand the principles according to which levers operate, one must understand the concept of *turning moment.* When an object rotates around an axis, the position, as well as the magnitude of the force which is turning it, is important. This can be tested very easily in the following way. Try to close an open door with one finger placed one centimetre from the hinges. This is very difficult. Repeat the procedure, placing the finger on the handle. The effort required is much less than in the former case. It is therefore evident that the further away from the axis the force is applied, the less force is necessary to perform a certain function. The perpendicular *distance* between the axis and the direction of the force on the one hand, and the *magnitude of the force* on the other, are inversely proportional to each other. The graph representing this relationship is a hyperbola (Chapter 1). The product of the two variables is constant and this constant is called the turning moment.

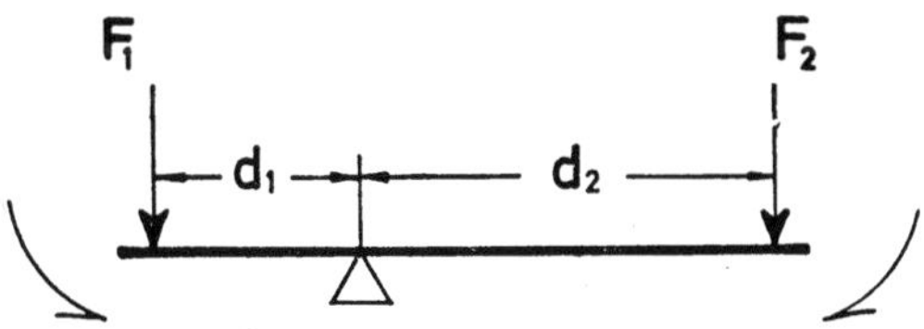

FIG. 17.5 No rotation takes place if the turning moments of two forces on each side of the fulcrum are the same.

When two forces are applied to the lever, one on each side of the fulcrum (FIG. 17.5), they tend to turn the lever in opposite directions. No rotation takes place if the turning moments of the two forces are equal, i.e.,

$$F_1 \times d_1 = F_2 \times d_2$$

where F_1 and F_2 are the magnitudes of the two forces and d_1 and d_2 the distances between the fulcrum and the forces. In this condition the lever is in equilibrium. Rotation away from this equilibrium position can be achieved by increasing one of the forces slightly. This can be illustrated by the operation of a see-saw. The two forces in this case are the weights of the two persons sitting at either end of the plank. The see-saw will remain in a horizontal position as long as the turning moments of the gravitational forces acting upon each person are the same. It is also known that the persons at each end of the plank do not necessarily have to have the same weight, provided that the heavier person sits nearer to the fulcrum than the lighter person. A large force can thus be balanced by a small force by simply adjusting the positions of the force in relation to the fulcrum, i.e., by making the two turning moments the same.

Problem 17.1. A boy of mass 40 kg, sits on one end of a see-saw at 1 m from the fulcrum. How far must a boy of 35 kg sit from the fulcrum in order to balance the heavier boy?

$$F_1 = 40 \text{ kg}$$
$$d_1 = 1 \text{ metre}$$
$$F_2 = 35 \text{ kg}$$
$$d_2 = ?$$
$$F_1 \times d_1 = F_2 \times d_2$$
$$\therefore \quad d_2 = \frac{F_1}{F_2} \times d_1 = \frac{40}{35} \times 1 = 1{,}14 \text{ metres}$$

The smaller boy must therefore sit further away from the fulcrum. The principle of the equilibrium of turning moments applies to all three types of levers. It must be noted that the larger the force, the closer it is to the fulcrum and the shorter the distance through which it moves, if rotation takes place (FIG. 17.6). The work performed by the two forces, i.e., force × distance moved, is the same for both forces (Chapter 3). The smaller force must therefore move through a greater distance than the larger force. Levers do not, therefore, lessen the amount of work, and this applies to other machines as well, since machines change only the magnitude and direction of forces, or the extent of movement caused by forces.

First and second class levers are generally used to lift a large load using a small effort. In first class levers, such as the see-saw and the weighing balance, the load and the effort work in the same direction. In second class levers they work in opposite directions. The wheelbarrow and the artery clamp are examples of second class levers. In the case of the wheelbarrow, the weight of the contents of the wheelbarrow is the load acting in the direction of the centre of the earth.

The effort is exerted by the muscles of the person pushing the

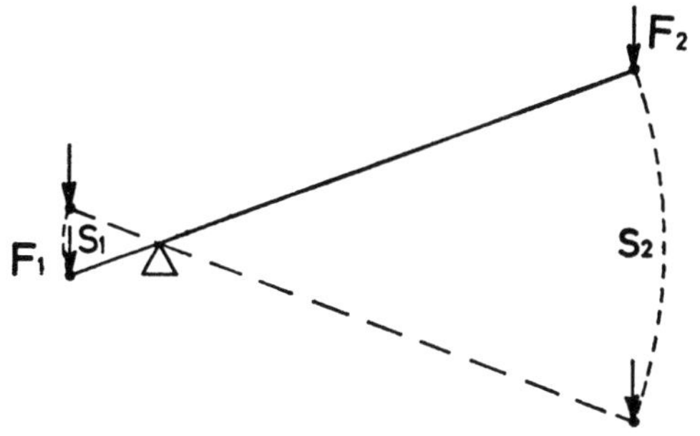

FIG. 17.6 The farther a force is from the fulcrum, the greater is the distance through which it moves.

wheelbarrow and this takes place in the opposite direction (FIG. 17.7). Third class levers are used when the available force is large enough, but the extent of movement caused by the force is insufficient. Because the work performed by the effort and the load must be the same, the effort must be greater than the load.

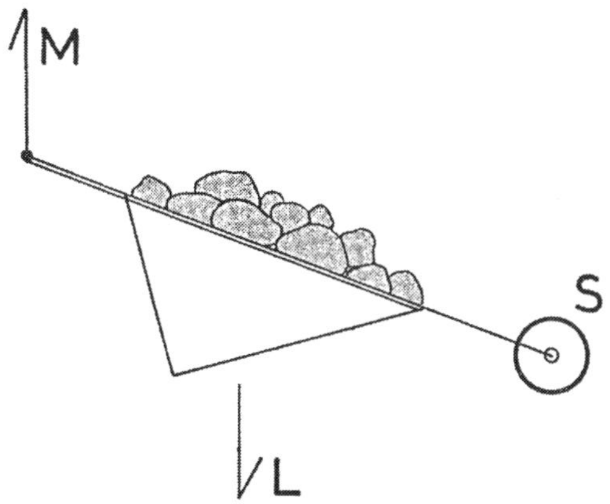

FIG. 17.7 Schematic representation of the forces acting upon a wheelbarrow. M = effort, L = load, S = fulcrum. This is an example of a second class lever.

Problem 17.2. A person must lift an object on which a force of 1 newton is exerted. The length of the lower arm is 0,30 m and the distance between the insertion of the biceps tendon and the elbow is 0,03 m.

(*a*) How great must be the force exerted by the muscle in order to balance the effect of the object?

(*b*) How much must the muscle shorten to allow the object to move through a distance of 0,5 m?

(*a*) $F_1 = 1$ newton $\qquad F_2 = ?$ newton

$d_1 = 0{,}30$ metre $\qquad d_2 = 0{,}03$ metre

$$F_1 \times d_1 = F_2 \times d_2$$

$$F_2 = \frac{F_1 \times d_1}{d_2}$$

$$= \frac{1 \times 0{,}30}{0{,}03} = 10 \text{ newtons}$$

The muscle exerts a force that is ten times greater than the force on the object.

The work performed by the two forces must be the same, i.e.,

(*b*) $F_1 \times S_1 = F_2 \times S_2$

$F_1 = 1$ newton $\qquad F_2 = 10$ newtons

$S_1 = 0{,}5$ metre $\qquad S_2 = ?$

$$S_2 = \frac{F_1 \times S_1}{F_2} = \frac{1 \times 0{,}5}{10}$$

$$= 0{,}05 \text{ metre}$$

The muscle, therefore, shortens a tenth of the distance through which the object is moved.

Pulleys

As already explained, a pulley is a circular disc that can turn around an axis situated at its centre. There are two types of pulleys, namely fixed and moveable pulleys. The first type, of which the knee-cap is an example, is used to change the direction of a force. Another example of the use of the fixed pulley is in the changing of the power of traction on a limb in certain types of fractures (FIG. 17.8). In some cases of fractures, for example, where the fractured ends overlap and the muscles contract spastically, proper healing is impossible, and the limb becomes shorter. To prevent this shortening, the limb is stretched during the whole period of recuperation. Generally the patient must lie prone, and the force of traction must be exerted horizontally. It is difficult to produce a horizontal force, but easy to obtain a vertical force, viz, the gravitational force which in fact is exerted on all objects and is most easily controlled by changing the size of objects. The direction of the force of traction can be changed from a vertical to a horizontal direction by means of a pulley.

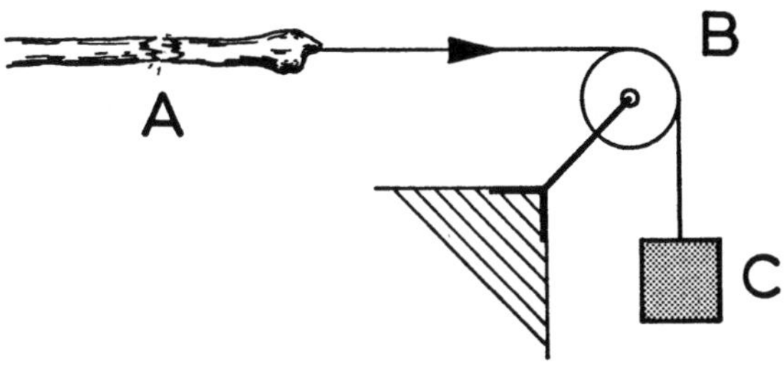

FIG. 17.8 Application of a pulley to produce an efficient and controllable traction in the case of a bone fracture. A = fracture, B = pulley, C = mass of the object on which the gravitational force is acting.

Moveable pulleys are used to reduce the effort in cases where objects are to be lifted. In the pulley shown in FIG. 17.9 the force on the object is divided between two sections of the rope, and traction is exerted on only one section. The person pulling the rope thus exerts half of the force acting upon the object.

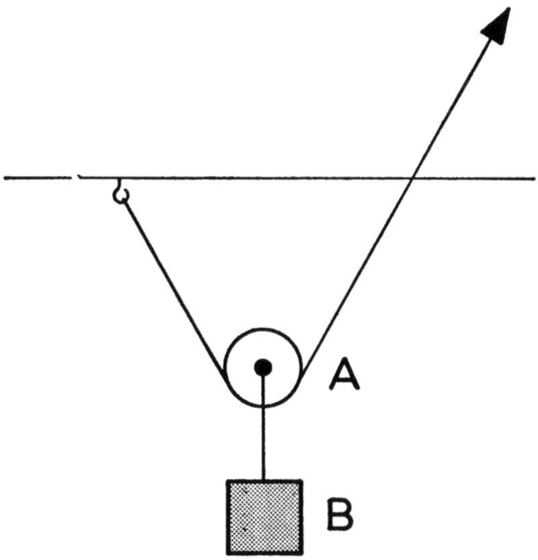

FIG. 17.9 Application of the pulley to reduce a force. A = pulley; B = load.

Body mechanics

We can exert forces by means of our bodies. These forces are produced by the muscles and are altered by the simple machines of the body in order to make them effective. The levers and pulleys of the body can be used in different ways. Wrong usage does not only mean wastage of muscular effort, but it can make some tasks impossible and can also damage the muscles, bones, tendons and joints. The science dealing with efficient usage of the machines of the body is called *body mechanics*.

The basic rule of body mechanics is that the *load must be as close to the fulcrum as possible.* This ensures a small turning moment of the load and this in turn means that a smaller effort is needed.

The easiest way of explaining this is to discuss the different ways in which a person is able to stand. If a person stands upright, the upper parts of the body rest on the lower parts, without showing any tendency to rotate. In this case the load is exerted at the fulcrum and its turning moment is zero. If a person sits or stands in a bent position, the gravitational force on the top parts of the body produces a turning moment about the fulcrum in the hip joint, and the muscles of the back must exert a greater force in order to cancel out this turning moment.

APPENDICES

Appendix A — The Elements and their Symbols

Appendix B — Examples of Examination Questions

APPENDIX A

The Elements and their Symbols

actinium (Ac)
alabamine (astatine, helvetium) (At)
alumin(i)um (Al)
americium (Am)
antimony (stibium) (Sb)
argon (Ar)
arsenic (As)
astatine (alabamine, helvetium (At)
barium (Ba)
berkelium (Bk)
beryllium (glucinum) Be)
bismuth (Bi)
boron (B)
bromin(e) (Br)
cadmium (Cd)
c(a)esium (Cs)
calcium (Ca)
californium (Cf)
carbon (C)
cerium (Ce)
cesium (Cs)
chlorin(e) (Cl)
chromium (Cr)
cobalt (Co)
columbium (niobium) (Cb)
copper (Cu)
curium (Cm)
dysprosium (Dy)
einsteinium (Es)
erbium (Er)
europium (Eu)
fermium (Fm)
florentium (illinium, promethium) (Pm)
fluorin(e) (F)
francium (virginium) (Fr)
gadolinium (Gd)
gallium (Ga)
germanium (Ge)
glucinum (beryllium) (Be)
gold (Au)
hafnium (Hf)
helium (He)
helvetium (alabamine, astatine) (At)
holmium (Ho)
hydrogen (H)
illinium (florentium, promethium) (Pm)
indium (In)
iodin(e) (I)
iridium (Ir)
iron (Fe)
krypton (Kr)
lanthanum (La)
lawrencium (Lw)
lead (Pb)
lithium (Li)
lutecium (Lu)
magnesium (Mg)
manganese (Mn)
masurium (technetium) (Tc)
mendelevium (Md)
mercury (quicksilver) (Hg)
molybdenum (Mo)
neodymium (Nd)
neon (Ne)
neptunium (Np)
nickel (Ni)
niobium (columbium) (Nb)
niton (radium emanation, radon) (Rn)
nitrogren (N)
nobelium (No)
osmium (Os)
oxygen (O)
palladium (Pd)
phosphorus (P)
platinum (Pt)
plutonium (Pu)
polonium (Po)
potassium (K)
praseodymium (Pr)
promethium (florentium, illinium) (Pm)
prot(o)actinium (Pa)
quicksilver (mercury) (Hg)
radium (Ra)
radium emanation (niton, radon) (Rn)
rhenium (Re)
rhodium (Rh)
rubidium (Rb)
ruthenium (Ru)
samarium (Sm)
scandium (Sc)
selenium (Se)
silicium (silicon) (Si)
silver (Ag)
sodium (Na)
stibium (antimony) (Sb)
strontium (Sr)
sulfur (sulphur) (S)
tantalum (Ta)
technetium (masurium) (Tc)
tellurium (Te)
terbium (Tb)
thallium (Tl)
thorium (Th)
thulium (Tm)
tin (Sn)
titanium (Ti)
tungsten (wolfram) (W)
uranium (U)
vanadium (V)
virginium (francium) (Fr)
wolfram (tungsten) (W)
xenon (Xe)
ytterbium (Yb)
yttrium (Y)
zinc (Zn)
zirconium (Zr)

PERIODIC TABLE OF THE ELEMENTS

1 H 1																	4 He 2
7 Li 3	9 Be 4											11 B 5	12 C 6	14 N 7	16 O 8	19 F 9	20 Ne 10
23 Na 11	24 Mg 12											27 Al 13	28 Si 14	31 P 15	32 S 16	35 Cl 17	40 Ar 18
39 K 19	40 Ca 20	45 Sc 21	48 Ti 22	51 V 23	52 Cr 24	55 Mn 25	56 Fe 26	59 Co 27	59 Ni 28	64 Cu 29	65 Zn 30	70 Ga 31	73 Ge 32	75 As 33	79 Se 34	80 Br 35	84 Kr 36
85 Rb 37	88 Sr 38	89 Y 39	91 Zr 40	93 Nb 41	96 Mo 42	98 Tc 43	101 Ru 44	103 Rh 45	106 Pd 46	108 Ag 47	112 Cd 48	115 In 49	119 Sn 50	122 Sb 51	128 Te 52	127 I 53	131 Xe 54
133 Cs 55	137 Ba 56	139 La 57 *	178 Hf 72	181 Ta 73	184 W 74	186 Re 75	190 Os 76	192 Ir 77	195 Pt 78	197 Au 79	201 Hg 80	204 Tl 81	207 Pb 82	209 Bi 83	210 Po 84	210 At 85	222 Rn 86
223 Fr 87	226 Ra 88	227 Ac 89 **	104														

*	140 Ce 58	141 Pr 59	144 Nd 60	147 Pm 61	150 Sm 62	152 Eu 63	157 Gd 64	159 Tb 65	163 Dy 66	165 Ho 67	167 Er 68	169 Tm 69	173 Yb 70	175 Lu 71
**	232 Th 90	231 Pa 91	238 U 92	237 Np 93	242 Pu 94	243 Am 95	247 Cm 96	247 Bk 97	249 Cf 98	254 Es 99	253 Fm 100	256 Md 101	254 No 102	257 Lw 103

ATOMIC MASS 65
Zn — SYMBOL
30 ATOMIC NUMBER

APPENDIX B

Examples of Examination Questions

1. Explain magnetic force and the application of this force in electromagnetic measuring instruments.
2. Discuss centrifugal force and one of its applications in biology.
3. Describe chemotaxic force.
4. Describe the graphical method used to calculate the resultant force when more than two forces act upon an object.
5. Explain what happens to Z in the equation $Z=1/y^2$ when
   (*a*) y doubles;
   (*b*) y decreases to one third of its value.
6. Describe the various steps to be followed when a physical law is formulated.
7. Draw the graphs representing a straight line, a hyperbola, a parabola and an inverse square relationship, and explain what can be deduced from these graphs.
8. Convert the following:
   (*a*) 15 km to mm
   (*b*) 45,3 m^2 to pm^2
   (*c*) 0,123 m^3 to μm^3
   (*d*) 1 750 mm to Mm.
9. Explain how an equation which has more than two variables is analysed.
10. Discuss the principles of an electrical cell.
11. Describe electrical force and its contribution to muscle contraction.
12. What are the carriers of electricity in:
    (*a*) metal conductors
    (*b*) fluids.
13. Which are the forces that make the movement of electrons in an orbit possible?
14. Describe the transmission of an action potential along a nerve fibre.
15. Describe the main constituents of an atom.
16. Sketch the equipotential lines produced by a dipole.
17. Discuss the origin of:
    (*a*) the resting membrane potential, and
    (*b*) the action potential, in living cells.

18. Describe, with the help of diagrams, the principles according to which:
    (*a*) the electrocardiograph, and
    (*b*) the synchronised defibrillator, function.
19. What is meant by complete heart block? How can it be treated by means of an electronic apparatus?
20. Define "good conductor".
21. Explain why it is possible for electrons to maintain a constant speed when they are travelling along a metallic conductor.
22. Explain how the resultant dipole in a nerve is effected by:
    (*a*) the diameter of individual fibres;
    (*b*) the number of fibres in a bundle.
23. Give a complete classification of the bio-electrical apparatuses with one example of each group that you mention.
24. Explain with the help of diagrams the principles according to which:
    (*a*) the oscilloscope, and
    (*b*) the synchronized pacemaker, function.
25. Discuss the SI-system of units. Give special attention to:
    (i) the structure of the system;
    (ii) the rules governing usage.
26. Explain the Kinetic theory.
27. Describe the various states in which matter exists.
28. Mention the physical factors which control the flow of fluids, gases and heat.
29. Discuss a few applications of the laws of flow in the body.
30. Why is the upper surface of water always horizontal?
31. Explain:
    (*a*) the mechanisms of lung function;
    (*b*) partial pressure.
32. Define hydromotive force (HMF).
33. Explain how a heart valve functions.
34. Define the concepts: solution, suspension, emulsion.
35. Describe the exchange of fluids between the blood and tissues in the capillaries.
36. What is meant by the terms: temperature and quantity of heat? Name the units concerned.
37. Describe the joule meter.
38. Explain, using the Gas Law, the mechanisms of respiration and the principles according to which artificial respirators function.
39. Describe the processes of diffusion and osmosis, and give examples where these processes participate in functions of the body.

40. How are the following pressures measured: atmospheric pressure, blood pressure, relative pressure and absolute pressure?

41. Discuss the factors which effect flow of fluid through a pipe with special reference to the cardio-vascular system.

42. Describe the temperature control mechanisms of the body.

43. Define wave-length and frequency of electromagnetic rays.

44. What is the speed of light?

45. Name the rays of the electro-magnetic spectrum in order of their wave-lengths.

46. Describe the relationship between speed of radiation, wave-length and frequency.

47. Explain how the following are produced:
    (*a*) light;
    (*b*) Röntgen rays (X-rays).

48. What is a laser?

49. How are rays reflected?

50. Explain how a ray is refracted when it passes from one material to another.

51. What is meant by the concepts:
    refractive power,
    isometric distance, and
    focus.

52. Explain the functional mechanism of a lens, and discuss the path of a ray through a concave lens and through a convex lens.

53. Describe what is meant by the concept accommodation of the eye.

54. Explain how the following function:
    (*a*) a magnifying glass;
    (*b*) a microscope.

55. List the precautionary regulations applicable when radioactive material is used.

56. Define half-life.

57. What is an isotope?

58. Describe the types of rays emitted by radioactive isotopes.

59. Define teletherapy?

60. What is a photomultiplier?

61. Define: pitch, quality and intensity of tone.

62. Explain the difference between a sound wave and a light wave.

63. What is the speed of sound.

64. Mention possible uses for ultrasonic apparatuses in medicine.

65. Give the range of audible frequencies.

66. Discuss the principles employed in diagnostic radioactive apparatuses.

67. Name the principles upon which correct posture of the body is based.

68. Give examples of the application of ultra-violet rays in biology.

69. Explain:
    (*a*) the functional mechanism of the eye;
    (*b*) optical defects;
    (*c*) the correction of optical defects by means of lenses.

70. Describe the simple machines of the body.

71. Mention the three classes of levers and discuss the principles according to which they function.